The

URANIUM
CLUB

The

URANIUM CLUB

UNEARTHING THE LOST RELICS OF THE NAZI NUCLEAR PROGRAM

MIRIAM E. HIEBERT

CHICAGO
REVIEW
PRESS

Copyright © 2023 by Miriam E. Hiebert
Foreword © 2023 by Timothy W. Koeth
All rights reserved
Published by Chicago Review Press Incorporated
814 North Franklin Street
Chicago, Illinois 60610
ISBN 978-1-64160-862-6

Library of Congress Control Number: 2023931789

Typesetting: Nord Compo

Every effort has been made to contact the copyright holders for the images that appear
in this book. The publisher would welcome information concerning any inadvertent
errors or omissions.

Printed in the United States of America
5 4 3 2 1

For J. J.

"The ultimate responsibility for our nation's policy rests on its citizens and they can discharge such responsibilities wisely only if they are informed."

—Henry Smyth

"I love physics with all my heart. It's a kind of personal love, as one has for a person to whom one is grateful for many things."

—Lise Meitner

CONTENTS

Part II: The Reactor Hitler Tried to Build

Part III: Gift of Ninninger

FOREWORD

THAT TWO OF THE URANIUM CUBES, artifacts of the development of nuclear physics during the Second World War, found their way to me feels in some ways like it was fated. For as long as I can remember, my love of science has been inextricably tied to my fascination with its history. This combined interest was initiated by Richard Rhodes's book, *The Making of the Atomic Bomb,* which was a gift from my uncle Charles when I was just ten years old. The whole story is told as a journey: the excitement, the thrill of the hunt, the fundamental secrets of nature. That captivated me. It also prepared me to appreciate the historical significance (or weight) of that first five-pound uranium cube.

But the part I loved best as a boy was the science detailed throughout that text, and my fascination with nuclear physics only expanded from there. I clearly remember sitting in my parents' basement, reading the theory of operation of an old civil defense Geiger counter that I had acquired when I was eleven years old. At that point I understood little of what I was reading, but I slowly learned by picking apart pieces of the concepts that I didn't understand and then going off and, one by one, figuring them out. Eventually I'd have a complete picture of how a given phenomenon really works and how it applies to the bigger system. Then, I would push the questioning a little further: *What would happen if we turned the high voltage up some more?* or *How could we make this smaller?* This was and still is my standard approach to learning. If you talk with any other scientist, they will probably describe a similar experience.

As I continued in my fascination with both the experimental details and history of nuclear physics, I found myself identifying with the scientists I was reading about—particularly Leo Szilard. It is hard to imagine the excitement that physicists working in the first half of the twentieth century must have felt as they used relatively simple experiments to yield each piece of the atomic puzzle and, one by one, put them into place. I wanted nothing more than to be able to join them in their work. I had found my tribe, so to speak—just seventy-five years too late.

But these legendary scientists were also human. Humans have the fascinating ability to be both deliberate and methodical as well as chaotic and emotional. Scientific advances often stem from the pure luck of phenomenal discovery. But for every stroke of fortune and genius, there are cautionary tales of the many instances when really smart people missed a discovery that passed right under their noses due to their own hubris or ignorance. Looking back across the history of science, these moments of victory and of loss become all the more obvious.

Today, the scientific research landscape looks very different than it did in the early days of atomic exploration, but still, so much of my understanding of how to approach my own research is based on the lessons I have learned from the lives and careers of scientists from the past. Chief among these lessons is to never do science—or any project, for that matter—alone. While it is often one man or woman's name that becomes attached to a new discovery, no scientist has ever accomplished any significant work without help. Without his friend and colleague Lise Meitner's interpretation of his results, Otto Hahn would have never received his Nobel Prize for the discovery of fission. I have seen this principle in action throughout my own life and career and have been fortunate to find myself among a cohort of diversely talented people who always make the product or outcome we are working toward far better than it would have been had I attempted to conduct it on my own. Sharing with and learning from the people around me has allowed my life and career to expand in ways that I never could have predicted—not the least of which has been the story that is contained in this book.

Being able to significantly contribute to a gap in the history of the dawn of the nuclear age—to hunt for, find and hold, study and analyze, and simply reflect on relics that played such a pivotal role in history—is a privilege my ten-year-old self could have only dreamed of. *The Uranium Club* details the foundations of not only the world's first instance of nuclear materials proliferation but also the spark that ignited the global nuclear enterprise that impacts every aspect of our modern lives.

The scientists in Rhodes's book faced innumerable complex and fascinating questions, and they suffered, they toiled, they did the grind to get the answers—to see something that no one else had seen before. As I have worked in my own laboratory, the most important lesson I have learned is that the mark of a true scientist is enjoying the pain of being wrong.

—Tim Koeth

1 | A CUBE APPEARS

TIM KOETH WAS MIDSTRIDE when the cell phone in his pocket rang, giving him a welcome excuse to pause his jog across the University of Maryland campus on a sweltering mid-Atlantic August evening. The fall term had not yet begun, and the throngs of students who would soon populate every inch of the sprawling campus still had not materialized. Catching his breath among the chalky colonial columns, Tim was alone. He looked at his phone and saw the name of one of his colleagues, Mary Dorman, the university's radiation safety officer, on the screen. Unsure of what Mary could want, Tim answered the call.

"Hi, Tim. Listen," began Mary. "We were cleaning out a laboratory in the geology department earlier today, and we found something—something I think you might want to see."

"Oh, really?" Tim replied, "What's that?"

"It would be better to just show you," Mary said. "Where are you? Can you meet me?"

They agreed to meet at a nearby campus parking lot, and Tim started in that direction.

Tim was intrigued, but not altogether surprised. This kind of cryptic phone call, seemingly full of mystery and intrigue, was not an unfamiliar occurrence in Tim's world. Tim is a physicist by training and trade, and by day he works as a professor sharing the wonders of the atom with his small army of devoted student researchers. But for Tim, physics isn't just a career. It's a way of life.

Tim's fascination with science—and with physics, in particular—began at a young age. When he was ten, his uncle gave him his first copy of *The Making*

1

of the Atomic Bomb by Richard Rhodes. He read it cover to cover more than once, turning the pages until the spine cracked in half.

By the time he was thirteen, Tim had begun his own scientific explorations at a little workbench laboratory that he had set up in his parents' basement. He had amassed a number of small instruments, including a couple of old Geiger counters, and after school one day he set out through the hallways of his former grade school building, radiation meter in hand, to see what he could discover. At first, the meter wasn't picking up much, but as he walked past the science classrooms, the sluggish clicks of the Geiger counter quickened. The clicks were fastest in the section of the hall that shared a wall with the science supply closet. He found a teacher who, seeing Tim's eagerness, agreed to let him investigate. Inside the closet, Tim discovered the source of the clicking. It was a small capsule attached to a thin rod, about the length of a pencil, that had been sitting in a cardboard tube jumbled among an assortment of beakers and Bunsen burners. Unsure what the object was, the teacher agreed to let Tim take it home.

Tim didn't know what exactly he had found, but he knew it was far more radioactive than anything he had collected before. The dial on his Geiger counter had hit its upper limit when he had brought it within several yards of the rod. On the car ride home in his mom's van, Tim held the rod away from his body, his arm outstretched, just to be safe. As soon as he got home, he ran down to the basement where he kept another meter, this one with a much higher upper detection limit. He flipped the meter on and held the rod above it; the dial immediately pegged to its highest setting.

Thrilled and terrified in equal measure, Tim considered what to do next. Several months prior, he had constructed a shielding container in his backyard, made from a metal barrel filled with concrete except for a four-inch hollow pipe down the center, nurturing the distant hope that maybe, just maybe, he would someday have a need for it. That wished-for day had arrived, and he rushed outside and dropped the rod into the center of the makeshift shielding vault. With the radiation risk now (somewhat) contained, Tim went inside to tell his parents what he had been up to.

The rod remained in its concrete housing overnight, and Tim went to school as usual the next day. When his mom picked him up that afternoon, her face was white as he climbed into the van.

"Did Dad get ahold of anyone?" Tim nervously asked.

"Oooooh yeah," his mother replied.

Arriving home, Tim was stunned to see the street blocked off by vans as men in hazmat suits rushed in and out of their house. Tim's father had spent the morning attempting to alert the authorities about what his son had found. Who exactly to contact had been unclear, but when he finally reached the right agency, the calvary had arrived in force.

Before Tim's mother left to get him, one of the suited-up men had informed her that even though the rod was outside, they were still picking up significant readings in the house. The only way forward, they had said, would be to dismantle the whole structure board by board, pack it in drums, and ship it to Washington State as radioactive waste. Hearing this, Tim, already the teacher, took it upon himself to calmly explain to this much older man that the house was not, in fact, contaminated—the meters were simply picking up radiation from the object outside through its shielding in the backyard. Eventually the men in suits left, taking the rod, which they believed to be an old medical radiation device, with them. The event had created such a fuss in the neighborhood that young Tim and his discovery were profiled in the local paper.

The adult, professorial version of Tim is not all that far removed from the starry-eyed boy tinkering away in his basement. He still has a basement lab, though it is now much bigger and better supplied. Much of his spare time is spent in this lair, working to replicate famous experiments in the history of science. And he is still constantly on the hunt for nuclear treasure.

But Tim is no longer alone in his fascinations. Over the years he has amassed a devoted following of former students, fellow tinkerers, and kindred treasure-hunting enthusiasts (myself included), who have all been drawn in by Tim's infectious fascination and indefatigable delight in the atom and all it can do—a strange society of nerds, sharing secrets and marveling at the world together.

Of all Tim's many roles, he is perhaps best known, in certain tight-knit circles, for his love of nuclear artifacts. Over the years he has gathered an impressive assortment of trinkets that played a role in, were witness to, or resulted from the nuclear race of World War II and the years that followed. These objects have found their way to Tim through sundry means. Some have been bought on the Internet or at auctions. The provenance of others is more eccentric: a couple of pounds of Trinitite* exchanged in a dusty New Mexico trailer park for a handle of vodka or an old medical device that was delivered via a two-thousand-mile relay from friend to friend across the country. But the most prized of his possessions found their way to Tim through sheer chance.

Tim had just completed his postdoctoral work at the University of Maryland when he was unexpectedly offered an opportunity to fulfill a lifelong dream. The Maryland University Training Reactor, a small nuclear reactor that had been installed at the university in the postwar nuclear boom, needed a new director. Offered the keys to his very own nuclear reactor, Tim leaped at the opportunity.

When he took charge of the reactor and the teaching program it supports, the facility that housed the two-story concrete pool with its glowing blue core was a mess. Old instruments and forty years of experimental equipment littered the floor and the surrounding rooms. It would take months to fully clean and organize the space.

One morning, Tim was wading through the chaos in one of the back rooms. On the far wall stood an old green metal cabinet that had once been used to store samples for experiments. He suddenly remembered that several years earlier he had been poking around in that same cabinet and had come across an interesting object. He wondered if it was still there and started pulling out the cabinet's contents. Most was trash—old sample vials and expired gloves. But tucked in the very back was an old forgotten white cardboard box, just a few inches wide. He opened the box, which was shockingly heavy given its size, to reveal a black metal cube, sitting nestled among yellowing tissue paper. The surface of the two-inch dark object was smooth and had been coated in some sort of shiny resin. Tim turned the box over in his hand and picked up the cube, which must have weighed over five pounds. As he turned the cube, he noticed that small notches, just a few millimeters deep, had been carved into four of the cube's edges.

Based on its weight and color, Tim guessed it was uranium—not an unexpected finding around a nuclear reactor. But as he examined the strange discovery, he could not shake the feeling that he had seen this cube somewhere before. He packed it up and arranged to have it transferred to his own lab space across the street for further study.

---

* The glassy material formed on the desert floor after the first nuclear test explosion.

Early one morning, several weeks later, Tim sat bolt upright in bed. In the night his mind had put together the pieces and he suddenly knew why uranium cube had felt so familiar. He rushed to his bookcase and pulled a small tattered red volume off a shelf. He flipped madly through the pages until he found what he was looking for. There, on page 170 of *Nuclear Physics* by Werner Heisenberg, was a picture of the last nuclear reactor experiment built in Germany during World War II.

In the image, the reactor core, an ominous chandelier of 664 cubes of metallic uranium suspended in a cylindrical lattice, hangs above a pool. The cubes were small, about two inches on a side, and would have each weighed about five pounds. In order to hold them in place while hanging, grooves had been filed into their edges so they could be wrapped in strands of aircraft cable that were cinched together above and below.

Tim let out a yelp as he realized that the cube he had found was somewhere in that grainy photo taken so many years ago. In all his years of collecting radioactive antiques, one of these cubes had never crossed his path. And now one had just fallen into his lap.

The discovery of the cube in the cabinet had happened almost two years before Tim started jogging his way toward the meeting location he had set with Mary. When he arrived at the appointed spot, he found Mary, a short woman with a massive amount of curly blonde hair, and an affinity for very high heels, standing outside her car, waiting with a huge smile on her face.

Tim, sweaty and excited, towered behind Mary as she opened the trunk of her car and reached in and pulled out an old, crinkled paper lunch bag. The small package was familiarly heavy.

Taking the parcel, Tim recognized the feel and weight of uranium metal and his heart leaped to his throat. He peered inside. The bag held a small black cube, with notches cut into two of its edges. Tim was stunned—finding one of these objects by chance felt like hitting the lottery, but two seemed beyond the realm of possibility.

"Do you know what it is?" asked Mary, grinning. From the look on Tim's face, she knew he did.

"I think so, but do *you*?" Tim replied, trying not to sound too eager.

"I have an idea. Do you want it?"

Bursting, Tim answered, "Absolutely, if no one else wants it."

"Happy birthday," said Mary with a smile.

Tim took his new prize and headed straight to his lab.

How this second cube had come to be in his possession, seventy years and thousands of miles separated from its original use, was a complete mystery. But unlike the first cube, whose journey from Nazi Germany to College Park, Maryland, was utterly undocumented, this one came with a clue.

When he opened the bag on a workbench in his lab, he found a torn piece of paper sitting at the bottom. Tim pulled out the small slip—it was covered in a fine black powder—and read the note written across it in faded pencil: "Taken from Germany from the reactor that Hitler tried to build. Gift of Ninninger."

2 | INTRODUCING ELEMENT 92

URANIUM, THE NINETY-SECOND ELEMENT on the periodic table, has an impressive scientific pedigree. It was first identified by Berlin chemist Martin Klaproth in 1798, in the waste rock of the Joachimsthal silver mines in Bavaria. Klaproth named the new element after the planet Uranus, which had recently been discovered by astronomers and which itself was named after the Greek god of the heavens. With few obvious uses, the new element would remain in relative obscurity for the next century, until 1896, when Henri Becquerel observed that uranium salts placed on an undeveloped glass photographic plate would cause the plate to become exposed, as if to a strong light. The source of this unexpected "radiation" remained a mystery until the publication of work conducted by Marie Curie and her husband, Pierre, on another element: radium.

Radioactivity, as the Curies named the phenomenon, is the result of an imbalance in the forces that hold the central nucleus of an atom together. The atomic nucleus is composed of positively charged protons and neutrally charged neutrons. Orbiting the nucleus are a third type of subatomic particle: negatively charged electrons. The number of protons in an atom of a given element remains constant—it is this number that gives an atom its identity. (Carbon has six protons, gold seventy-nine, and uranium ninety-two.)

Generally, one finds about as many neutrons as protons in an atom's nucleus, but that's not a hard-and-fast rule. For example, some carbon atoms have six neutrons while others have seven or eight. Atoms of the same element but with different numbers of neutrons are called *isotopes*.

7

Neutrons and protons have essentially identical atomic masses, and different isotopes of atoms are classified by their total atomic mass: the number of protons plus the number of neutrons (electrons have almost no mass and aren't included). Thus, the three isotopes of carbon are ^{12}C, ^{13}C, and ^{14}C.

Within the nucleus of an atom, the positive charges of the protons cause them to repel one another, like matching poles on a magnet. The presence of neutrons, which are attracted to and sit between protons within the nucleus, allows the forces in the atom to remain balanced so that the nucleus remains stable. But this balance can be thrown off by the presence of additional neutrons, making the nuclei of some isotopes less stable than others.

Radioactivity occurs when these less-stable isotopes expel a particle or energy wave in order to become more stable, a process referred to as *decay*. This can be precipitated in some elements when the atom is hit with a particle or energy wave, but for certain isotopes of some elements it can also happen spontaneously. There are several different forms of radioactivity, and the type of radiation emitted depends on the isotope in question, and the mechanism behind its decay.

Not all isotopes are created equal. The percentage of the total number of atoms of a given element existing as a particular isotope can be extrapolated from empirical measurements and vary considerably. Returning to carbon, for example, ^{12}C makes up the majority (98.9 percent) of carbon in the universe, while ^{13}C makes up 1.1 percent of all carbon, and only one in one trillion carbon atoms is ^{14}C.

Despite its scarcity, ^{14}C is, radioactively speaking, the most interesting. The extra neutron renders the nucleus of these atoms unstable, unlike their lighter, more abundant counterparts. More important, ^{14}C spontaneously decays at a well-defined rate. Every 5,730 years, half the ^{14}C present in a sample will have undergone radioactive decay to the stable isotope of nitrogen-14 (^{14}N).* This period, over which half the original number of unstable atoms will have decayed to obtain a stable state is referred to as *half-life*. The radioactive isotopes of some elements have half-lives that last only fractions of a second, while others can encompass millennia.

Uranium is a much larger atom than carbon, and with its larger size comes a greater degree of instability within its voluminous nucleus. Because of its size, many different isotopes of uranium are possible, some found in trace proportions

---

* This is the functioning principle behind carbon dating.

in nature and some existing only within the confines of a laboratory. For the scientists studying nuclear fission and radioactivity in the early twentieth century, only the two isotopes most often found in nature were of importance. Approximately 99.27 percent of all the uranium on Earth is ^{238}U, while the remaining 0.72 percent is ^{235}U. Unlike carbon, for which some of the isotopes are stable and some are not, all isotopes of uranium are at least somewhat radioactive. The larger isotope, ^{238}U, is stable compared to its lighter counterpart, with a half-life of 4.5 billion years. The less abundant isotope, ^{235}U, has a comparatively much shorter half-life of only about seven hundred million years.

In addition to more "normal" radioactive decay, the extreme size of uranium's nucleus allows it to undergo a different type of breakdown unavailable to the vast majority of the elements on the periodic table: fission. Under certain circumstances, a free neutron, one that is not attached to the nucleus of any atom, can hit the nucleus of an atom of ^{235}U. This collision causes the uranium nucleus to break apart into (usually) two pieces, releasing energy and ejecting a few extra neutrons in the process. These freed neutrons can then collide with other neighboring atoms of ^{235}U, causing the process to repeat, and forming a "chain reaction."

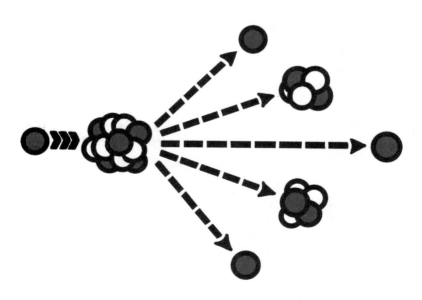

Fission diagram. *Created by Benjamin Krapohl.*

This ability to participate in a self-sustaining chain reaction of fission events makes ^{235}U what is referred to as a *fissile* material. It is, in fact, the only naturally occurring fissile material in existence in any meaningful quantity on Earth. Unlike the minority isotope ^{235}U, the majority isotope of uranium, ^{238}U, cannot be used to initiate or sustain this type of chain reaction.*

Uranium is a relatively abundant element within the earth's crust—more abundant than gold or silver. It is found in at least trace amounts nearly everywhere on Earth, with major deposits of uranium ore and uranium oxide found in Australia, Kazakhstan, and central Africa. There is also an enormous amount of uranium dissolved in the world's oceans, though the concentrations are too dilute to be dangerous.

Uranium, in its purest metallic form, is remarkably dense. A block of uranium metal weighs nearly 70 percent more than a block of lead of the same size. People who have had an opportunity to pick up one of Tim's uranium cubes never expect the enormous weight of such a small object—most people must use two hands.

In nature, uranium is seldom found in a metallic form but is usually part of some sort of oxide or other chemical compound. These compounds often take the form of minerals or powders, and range in color from an extremely saturated black to bright white to lemon-yellow. Thanks to this wide range of colors, uranium compounds found their way into many commercial products as a colorant in the early 1900s.

Vaseline glass—the pressed glass plates, bowls, and ashtrays that still adorn many a grandparent's living room today—was widely manufactured and became popular during the Great Depression, hence the moniker Depression glass, as it is sometimes called. The strange sickly green color that is characteristic of many of these objects stems from small amounts of uranium oxide added into the molten glass during manufacturing.

Uranium oxide also found its way into glazes for ceramics; once fired, it turns a tomato-orange color. This color was popular during the early twentieth century, and staged a comeback after World War II, regaining popularity in the 1950s and '60s. While no longer widely manufactured, both the glass and ceramics are now popular collectibles and can often be found in antique stores.

---

* Though it can be used to breed another artificially produced fissile material—plutonium. More on that later.

But its role in the development of the atomic bomb in 1940s is what really cemented uranium's celebrity status. Few elements on the periodic table can boast the kind of exposure that uranium has received over the last several decades; it has been featured prominently in multitudes of Hollywood movies, from cult classics like *Dr. Strangelove* to the more recent HBO miniseries *Chernobyl.*

Uranium also often finds itself the topic of discussion on the news, as experts weigh in on how much of it certain countries have, how much they should be allowed to possess, and what exactly they are doing with it. But despite its high profile and relative ubiquity, it is an element that few people knowingly encounter in substantial proportions in their day-to-day lives. Unlike uranium glass, a five-pound cube of uranium is never going to turn up on the shelves of your local antique shop. The story of how these objects found their way to Tim Koeth is interwoven with the history of World War II and the race between Germany and the United States for nuclear supremacy.

3 | A BRIEF HISTORY OF FISSION

AUSTRIAN PHYSICIST LISE MEITNER was spending her winter holidays in the small town of Kungälv in western Sweden in late December 1938 when she received a letter from her colleague and close friend German chemist Otto Hahn. Until July of that year, Meitner had been living in Berlin, working with Hahn and their colleague, Fritz Strassmann, at the famed Kaiser Wilhelm Institute for Chemistry. At a time when most scientific institutions were at best apathetic, and at worst openly hostile to the inclusion of women in science, the small, serious Meitner had risen to a position of prominence at one of the premier scientific institutions in the world. Considered by many to be one of the great experimentalists of her time, fellow German physicist Albert Einstein would often refer to her fondly as "our Marie Curie."

Though the Meitner family was of Austrian-Jewish descent, Meitner had not been raised in the Jewish religion or community during her childhood in Vienna and had been baptized as a Protestant in 1908. But in bloodline-obsessed Nazi Germany, Meitner's Jewish ancestry was more than enough to render her a Jew in the eyes of the Third Reich.

For a time, in the early days of the Nazi regime, Meitner's Austrian citizenship afforded her a thin veil of protection. But after Germany annexed Austria in March 1938, Meitner was suddenly subject to the same ruinous restrictions that had been placed on German Jews, which included exclusion from academic posts. Without a secure position, Meitner's livelihood and potentially her life were suddenly in jeopardy. Leaving Germany quickly became her only option. A network of scientists across Europe sprang into action to aid their

13

friend, helping to secure permission for Meitner to leave Germany and to find her an academic position in Stockholm once she escaped. On July 13, 1938, Meitner said goodbye to Hahn, who gave her his mother's diamond ring to pawn, before heading to Germany's border with Holland.

Though her paperwork was in order, when Meitner arrived at the border she was unsure if the Nazi border agent would allow her to pass. But a misogynistic quirk of German grammar worked in Meitner's favor: the agent who glanced at Meitner's passport saw her name listed as "Frau Professor Meitner" and assumed, incorrectly, that Meitner was only a professor's wife, rather than a professor herself. Meitner was allowed to enter Holland, and while she would visit Hahn in Berlin years later, she would never again call Germany home.

After helping his friend escape, Otto Hahn remained in Germany and continued his work with Strassmann at the Kaiser Wilhelm Institute. Following the discovery of the neutron in 1932, Hahn and Strassmann had been conducting experiments in which these neutrally charged particles were shot into the nucleus of atoms as a means of probing the forces at play within the subatomic structure. Specifically, one experiment involved firing neutrons into the nuclei of uranium atoms. They had noticed that one of the products of this bombardment was the seemingly spontaneous presence of the much lighter element barium. Neither Hahn nor Strassmann could make any sense of this result, and so they wrote to Meitner, explaining what they were seeing and asking for her help.

At first Meitner was as baffled as her colleagues by the findings. Hahn's letter had arrived just before Christmas, and Meitner's consideration of the matter was interrupted by the arrival of her nephew, Otto Frisch, who also happened to be a physicist working at the Niels Bohr Institute in Copenhagen.

One afternoon in the first days of January 1939 Meitner and Frisch set out for a walk in the hills around Kungälv. The sky was clear, and the sunlight sparkled on several inches of snow as the pair made their way through the trees. Frisch walked in skis, lifting him several inches above his aunt, who, sinking through the snow cover, was following on foot. As they walked, Meitner told Frisch about the letter she had received from Hahn. She detailed the experiments that were being conducted in Germany, and her puzzlement over the results. As they talked the matter over, the indisputable facts underpinning the experiment became increasingly clear, until quite suddenly, in a jolt of inspiration often ascribed to scientists but rarely experienced, Meitner understood what was happening. She sat down on a nearby fallen log, pulled off her gloves, and rooted around in her coat pockets for a pencil and a scrap of paper. Frisch sat down next to her, and within ten minutes the pair had quickly worked out the math to describe what Hahn and Strassmann had observed.

Upon striking the uranium nucleus, the neutron had split the atom apart, forming several smaller atoms and a few neutrons, and also releasing a large amount of energy. Using Einstein's famous equation, $E = mc^2$, which relates mass to energy, Meitner and Frisch calculated that the masses of the expelled particles, when combined with the energy released, all added up to the original mass of the uranium atom. There, in the freezing cold of the glittering Swedish forest, Meitner and Frisch had just described what would become the most influential scientific development of the twentieth century: fission.

Meitner and Frisch continued their collaboration into the new year, publishing their results in the scientific journal *Nature*. Hahn and Strassmann also published their results, but did not mention Meitner's role in the discovery. Though Hahn would be awarded the 1944 Nobel Prize in Chemistry for the discovery of fission, Meitner would go largely unrecognized for her essential contribution.*

Word of the new discovery quickly began to spread. Otto Frisch brought news of fission directly to his boss, Niels Bohr, in Copenhagen. Bohr, a Danish physicist who became a giant in the field, was known worldwide, both for his own contributions to fundamental physics as well as for mentoring and spurring on the discoveries of countless other scientists. When Bohr arrived in Washington, DC, for a meeting of the American Physical Society later that month, he presented the work of Hahn and Strassmann, along with Meitner and Frisch's interpretation to the American scientists. Stunned by what they were hearing, several of the physicists in attendance abruptly left the auditorium and rushed to their laboratories before Bohr had even finished speaking.

What most physicists the world over realized immediately was that fission had implications beyond probing the inner workings of atoms. They understood that, in addition to smaller atoms, a likely product of the fission of a uranium nucleus

---

* Unlike her nephew, who would find himself at Los Alamos working on the world's first atomic weapon in just a few years' time, beyond her discovery of fission, Meitner declined any further involvement in the extension of her discovery into its inevitable terrible end. Her tombstone in Hampshire, England, where she died in 1968, reads: A PHYSICIST WHO NEVER LOST HER HUMANITY.

was one or more neutrons. These newly freed neutrons would be sent flying out of the broken atom and would be liable to collide with another neighboring uranium atom. If the free neutron struck the nucleus of another atom, it could cause another fission event, which would, in turn, release more neutrons, creating a chain reaction of self-perpetuating fissions. With each fission event comes the release of energy. While this burst of energy is relatively small for a single fission, a chain reaction of the hundreds of millions of possible fissions that could occur within even a small amount of uranium could generate enough energy to power an entire city—or create a bomb powerful enough to utterly destroy one.

The possibility of harnessing this power to generate energy, or, as the probability of a second world war increased, to build a weapon, was not lost on either side of the Atlantic. But it was the community of immigrant scientists who had recently escaped Europe and understood the grave danger posed by a new superweapon in Nazi hands who were the first to ring the alarm bells.

Recognizing the immense destructive potential of nuclear power and the need for the United States to act immediately in pursuit of harnessing it, the Hungarian American physicist Leo Szilard wrote the famous letter, signed by Albert Einstein, that was sent to President Franklin D. Roosevelt on August 2, 1939, only eight months after the publication of Meitner's findings. This letter urged Roosevelt to take seriously the potential implications of nuclear power and nuclear weapons and warned of the possibility that German scientists might try to produce such weapons themselves. Little information was coming out of Germany after all formal communications had been abruptly cut off at the onset of the war in Europe, but many assumed that Germany, which had been at the center of modern physics for the last half century, was indeed launching a nuclear weapons program.

Though not entirely convinced of the urgency of the matter, Roosevelt heeded Einstein and Szilard's warning and authorized the formation of the Advisory Committee on Uranium, a modest group with members from the army, navy, and the National Bureau of Standards, who met occasionally to discuss progress on the uranium problem. The scope of the committee's work remained fairly limited until April 1940, when new intelligence obtained from Europe seemed to suggest that scientists at the Kaiser Wilhelm Institute for Physics in Berlin were working on a much larger nuclear effort. The Hungarian's worry, it seemed, had been entirely justified. In response, the Advisory Committee on Uranium was elevated to the status of a subcommittee of the new National Defense Research Committee (NDRC) under the oversight of Vannevar Bush. The project's new standing also came with funding, and by the autumn of 1941 the program had extended to include sixteen individual research projects being carried out at a number of universities across the country, and a total annual budget of about $300,000.

Much of the early work on nuclear fission in the United States was conducted at Columbia University in New York City. This early charge was led by Enrico Fermi, the famed Italian physicist who had emigrated to America in January 1939. A slight man with a rapidly receding hairline, Fermi had gained unprecedented prominence in the Italian physics community. But because his wife, Laura, was Jewish, Fermi began to fear for the safety of his family amid the growing antisemitic sentiment and violence sweeping Fascist Europe. When Fermi was awarded the Nobel Prize in Physics in 1938 for his work on artificial radioactive elements, he traveled with his family to Stockholm for the ceremony and then, instead of returning to Italy, fled to the United States. Upon his arrival in America, the newly minted Nobel laureate was offered prominent positions at several American universities and accepted an offer to work at Columbia University in New York. Along with Szilard and with University of Chicago graduate student Herbert Anderson, Fermi began working on the design for a uranium nuclear reactor that could achieve a self-sustaining chain reaction.

Fermi's experiments at Columbia became a critical first milestone in determining the feasibility of a nuclear weapons program. Building a critical reactor pile would serve as proof of principle that a nuclear chain reaction was, in fact, possible and controllable—both necessary prerequisites for the development of an atom bomb.

Within a matter of months, the nuclear project had outgrown its position as a subcommittee of the NDRC and was instead put alongside its former parent committee under the overarching new Office of Scientific Research and Development (OSRD). On December 18, 1941, just eleven days after the Japanese attack on the US naval base at Pearl Harbor and America's entrance into the war, Vannevar Bush, now head of the OSRD, convened a meeting to discuss the acceleration of work toward an atomic weapon. Bush and his colleagues understood that any hope of harnessing nuclear power in the form of a weapon would require an all-out effort by a dedicated team. A special district was created to oversee and organize this work, and control of the project was formally transferred to the US Army Corps of Engineers.

In the early days of the new district, its creation and organization were led by Colonel James C. Marshall, who set up a headquarters for the project at 270 Broadway, in Manhattan. Marshall appointed Lieutenant Colonel Kenneth Nichols as his deputy. But by the end of the summer in 1942 Bush had grown impatient with the slow pace of the project's development and a change in leadership was suggested.

When then-colonel Leslie Groves of the US Army Corps of Engineers was initially approached about leading the new top-secret weapons program, he had just finished managing the construction of the new Pentagon building outside Washington, DC. Groves had been eager for his next assignment to take him overseas to an active military theater, so when he was told of his new appointment in September of 1942, his "initial reaction was one of extreme disappointment." His new assignment would not only keep him in Washington; the scope of the endeavor was expected to be relatively small, not costing more than $100 million. Groves had been made aware in passing of the planned nuclear project, and he recalled: "What little I knew of the project had not particularly impressed me, and if I had known the complete picture I would have been still less impressed." But, despite his reluctance, Groves was given little choice in the matter, and in September 1942 he formally took control of the program.

Enrico Fermi and his team were by this point already well on their way. In November 1942 they had begun construction of a major reactor experiment at the University of Chicago. Their lab was constructed in a basement squash court, tucked under the stadium seats of Stagg Field. Much larger than the previous experiments Fermi had been working on in New York, the enormous pile, named Chicago Pile-1 (or CP-1), comprised 385 tons of stacked graphite blocks. Interspersed between the graphite or embedded in the bricks themselves were over six tons of uranium metal and forty tons of pressed uranium oxide powder formed into cylinders and "pseudospheres."*

Fermi's use of graphite in his experiment was notable. In addition to uranium, every nuclear reactor also needs a moderator—a material that will slow the movement of the neutrons flying out of fission events enough that they are able to interact with other uranium atoms. There are only a handful of materials with this unique ability to slow neutrons without absorbing them. Heavy water—water (H_2O) in which the hydrogen atoms have been replaced

---

* A sort of odd double-sided cone shape that was meant to mathematically approximate spheres while being easier to machine.

by their heavier isotope deuterium to make D_2O, has this ability. But heavy water is complicated to use, and time consuming and expensive to produce.

Fermi and his team knew that carbon-rich materials, like graphite, should be similarly effective at slowing neutrons. Graphite was also inexpensive and easily produced, and the solid format made its use in pile construction much less complex—a seemingly perfect alternative. But early tests using graphite as a moderator showed troubling results. Instead of being slowed, the neutrons in these experiments were disappearing. Rather than scrap the idea of using graphite, the Manhattan Project scientists instead decided to look into why the material was not working as predicted. The culprit, they came to find, was a minute amount of boron, which is a commonly found contaminant in natural graphite sources. Boron is an avaricious absorber of neutrons, and the small amount in the graphite was gobbling up the neutrons set loose in the pile, thereby poisoning the experiment's results. Diligent processing of the graphite to remove the boron was all that was required to correct the issue before Fermi and his team set to work building their reactor pile with dark carbon bricks.

Lithograph of Chicago Pile 1. *Wikimedia Commons*

Construction of the massive pile took weeks, with two teams working around the clock in twelve-hour shifts. On December 2, 1942, before it was even completed, Fermi and his fellow scientists gathered around the control panel situated on a walkway overlooking the court below, the dark, radiation-laced mountain of graphite rising before them. Very slowly, Fermi began to turn the dial that would draw the cadmium control rods (the brakes for an accelerating nuclear reaction) out of the pile, and the neutrons began to fly. While inserted within the pile, control rods absorb more neutrons than the pile can create, but as they are withdrawn from the pile's center, the number of neutrons is able to increase. Each time Fermi turned the knob, extracting the control rods farther, the scientists watched in silence as the thin red line being drawn on the rotating chart recorder, like a painfully slow seismograph, tended gradually upward before leveling off. The entire process took several hours, with the scientists pausing midway to break for lunch. Finally, in the afternoon, Fermi turned the knob one last increment further. Instead of leveling off, the pen on the chart recorder began to sweep sharply upward, indicating a steadily increasing number of neutrons; CP-1 had become the world's first functioning nuclear reactor. After only a few moments, Fermi dropped the control rods back into the core, and the number of neutrons plummeted with them.

Among those present for this historic event was Arthur Compton. Nearly fifty, tall and dark with a narrow black mustache, Compton had been appointed by Vannevar Bush in 1941 to head a special committee tasked with tracking and reporting on the progress of the uranium program. In the wake of Fermi's successful demonstration of criticality, the University of Chicago physics professor famously telephoned his friend and supervisor at the NDRC, James B. Conant: "You'll be interested to know that the Italian navigator has just landed in the new world."

The nuclear age had begun.

In light of Fermi's proof of the possibility of a self-sustaining nuclear chain reaction, harnessing this source of seemingly unlimited power took on a new urgency. Under the oversight of recently promoted Brigadier General Leslie R. Groves, the newly named Manhattan Project rapidly expanded.

To carry out the work, three major sites were constructed across the country. Oak Ridge, Tennessee, became the site of the world's second functioning nuclear reactor, X-10, which was used to develop a method for the production of plutonium for use in weapons. Eventually, full-scale production of plutonium was moved to the project's Hanford site in Washington State. The Y-12 plant, also at Oak Ridge, worked to separate the two major isotopes of uranium, extracting the fissile minority isotope ^{235}U, which, when isolated from the heavier ^{238}U, could be used as fuel for an atomic weapon.

The design and engineering work for the construction of the atomic bomb was carried out at Los Alamos in New Mexico. Located about forty miles northwest of Santa Fe, the remote site had once been a boys' boarding school. The isolated location made it the perfect spot for the US Army Corps of Engineers to set up an ad hoc city in the desert, complete with a library, a movie theater, a radio station, and even stables. The scientists who were recruited to work at Los Alamos were required to move to the site with only their immediate family, telling no one else where they were going or how long they would be away. A post office box address in Santa Fe (PO Box 1663) was their only connection to the outside world as the scientists and engineers worked around the clock developing the designs for a functioning nuclear weapon.

On July 16, 1945, Los Alamos scientists tested the world's first atomic bomb, code-named the Gadget, at the Alamogordo Bombing and Gunnery Range in the Jornada del Muerto desert of New Mexico. This site, located about two hundred miles south of Los Alamos, was nicknamed "Trinity" by the director of Los Alamos Laboratory, Robert Oppenheimer. He chose the name in part in reference to the fourteenth of John Donne's *Holy Sonnets*, which begins: "Batter my heart, three-person'd God." The name stuck.

The night before the test, the device, a gray metal sphere about five feet in diameter, covered haphazardly in splotches of tape, was driven out to the site on a truck bed; the plutonium core was delivered by car. The core was carefully inserted into the implosion shell, and the whole apparatus was hoisted to the top of a hundred-foot steel tower.

Groves, Conant, and Bush watched the test together from the base camp set up ten miles away from the tower. Oppenheimer observed the blast from a bunker from half that distance. As they awaited the detonation, the observers had been "told to lie face down on the ground, with [their] feet toward the blast, to close [their] eyes, and to cover [their] eyes with [their] hands as the countdown approached zero."

Nervous silence filled the desert. No one was certain what to expect from the explosion. In fact, they weren't even sure the device was going to work at all—they had even built, though they did not use, a containment module to try to prevent the loss of any precious plutonium metal in case the experiment "fizzled." But these precautions proved unnecessary, and at precisely 05:29:45 in the morning, following a delay due to inclement weather,* the Gadget was detonated.

A larger group of scientists and engineers had assembled farther away to watch the blast from the top of Compañia Hill, about twenty miles northwest of the tower. As they silently watched the horizon through sunglasses layered under welder's goggles, the first indication of Gadget's successful detonation was the sudden illumination of the early-morning desert landscape around them with an eerie purple light several times brighter than the midday sun. The intense flash was followed a few seconds later by a rush of hot air, as the immense heat from the explosion, tempered by its travels across the sand, enveloped them. The explosion was so hot that the sand coating the desert immediately around the detonation site was liquefied, turning into a greenish-brown glassy material now called Trinitite and prized by collectors. Finally came the sound, a tremendous roar that was felt as much as heard, as an umbrella of fire, molten sand, and radioactive debris stretched skyward.

The Trinity explosion, which released an estimated eighteen kilotons of force,† demonstrated the feasibility of nuclear weapons—they could be created, and their explosive potential was everything that physicists had feared when they had first heard of Meitner's discovery six and a half years earlier.

In the intervening years since the war began, little substantive information had come out of Germany. But, throughout the Manhattan Project's effort, it had been widely accepted as truth that, given their initial head start, the

---

* According to legend, General Groves, only somewhat in jest, threatened to hang the meteorologist working with the group, John Hubbard, if his predictions about the weather clearing up at dawn proved incorrect.
† The units of kilotons and eventually megatons are used to describe the force of atomic and thermonuclear weapons in equivalent weights of TNT (i.e., dynamite). Fifteen kilotons of force is roughly equal to fifteen thousand pounds of TNT.

Germans' research program had to have kept pace with American progress, if not exceeded it. It was, in large part, this fear of a nuclear Germany that had motivated the high priority and breakneck pace of the Manhattan Project. But by the time of the Trinity explosion, the Third Reich had already surrendered, leaving Japan as the only remaining adversary in the war, and nothing close to what the scientists had witnessed in New Mexico had ever appeared in Hitler's arsenal.

PART I

TAKEN FROM GERMANY

Colonel Pash, Agents Leonard and Beatson, and Mascot "Alsos"; all set for Paris operation. *National Archives II, College Park, RG 165, Box 161, Alsos Mission Album*

4 | THE LAWYER: JOHN LANSDALE JR.

FOR THE DAUGHTERS OF John Lansdale Jr., the heavy metal cube that sat in their father's office was as innocuous as any of the other household knickknacks that filled their childhood home in the meticulously manicured community of Shaker Heights, Ohio. Growing up, Helen, Chloe, Mary, Metta, and Sally knew that their father had been involved in the war, and that the small but heavy black cube, encased in a thick cylinder of Lucite plastic, which periodically shifted from its position among family photographs on his large desk to holding back the open blinds on a corner of the windowsill, was made of uranium. But they knew little about where it had come from, or how their father had acquired it.

Today, the cube sits on a bookshelf in Omaha, Nebraska, in the living room of Sally Lansdale, the youngest of the five sisters. Over time, the Lucite encasement, which enlarges the object from a two-inch cube to a plastic puck with the diameter of a coffee can, has grown yellow and opaque, and a large chunk of one face has broken off, exposing the black metal below. The bare cube face is covered in cavities and voids caused by the rushed casting process, though it lacks the same notches carved into any of its edges like both of Tim's cubes. The whole object sitting on Sally's bookcase has been wrapped in a thin flexible sheet of lead, presumably to prevent any unwanted radiation from escaping into Sally's living room. Despite this additional shielding layer, other people tend to be much more wary of the cube than the Lansdale daughters, who grew up with it. "After all," Sally laughed, "Dad lived to be ninety-two. How dangerous could it be?"

Born in California but raised in Texas, John Lansdale Jr. grew up in a military family—a Lansdale had fought in every American war since the revolution.

27

Young John followed suit, and after graduating from the Virginia Military Institute (VMI) in 1933, he received a commission in the US Army Reserve as a second lieutenant with assignment to field artillery. While serving in the reserves, Lansdale attended law school at Harvard University, traveling back home to Texas for his two weeks of active duty during the summers of 1934 and 1935.

These trips afforded Lansdale the opportunity to visit his childhood sweetheart, Metta Virginia Tomlinson. Metta liked to say that she and John "had never really met; he was just always around"—both had grown up in the Houston area, and their families had attended the same church. Lansdale graduated from Harvard in the spring of 1936, and on June 17 of that year he and Metta were married.

The newlyweds moved to Cleveland, Ohio, where Lansdale began work at the corporate law firm of Squire, Sanders & Dempsey. Their first two daughters, Helen and Chloe, were born in the following two years.

A man of short stature with a kind-looking, bespectacled oval face and a penchant for red bow ties, Lansdale had an unassuming appearance that masked an enormous amount of energy and intellectual curiosity. As a newcomer to the corporate law firm, Lansdale worked long hours, spending nearly every day at the courthouse filing paperwork and making arguments to defend various corporate clients from a near constant string of lawsuits. The unending work and ever-moving targets of these cases gave Lansdale the opportunity to cultivate an ability to think quickly and pursue multiple goals at once. He also discovered through his time in court that he had quite a "talent for fakery."

As German aggression ignited a war in Europe, and Japan, by signing on to the Tripartite Pact with Germany and Italy, extended that war into the Pacific, the US military, recognizing the near inevitability of American involvement, began taking steps to increase their numbers. Part of this effort involved offering active-duty positions to reserve officers. With a new wife, a mortgage, and two small children, Lansdale initially evaded these increasingly insistent offers. But eventually, knowing that those who willingly volunteered for service were likely to have more say in their appointment than conscripts, Lansdale decided to accept.

In May 1941, as he was preparing for active service in what he assumed would be an appointment as a field artillery instructor at Fort Sill, Oklahoma, Lansdale received a letter from VMI classmate Frank McCarthy, who was serving as the secretary to the general staff of the War Department and recruiting additional personnel for the Military Intelligence Division. McCarthy suggested that Lansdale consider making a request to join his division in Washington, DC, instead. Lansdale decided to take his friend's advice, and on June 10, 1941, he received orders to report for active duty as a first lieutenant under General George Veazey Strong, the deputy chief of staff for intelligence (G-2) for the War Department. His initial appointment was to the Investigation Branch, Counterintelligence Group.

The young Lansdale family packed up their belongings and moved from their house in Ohio to the upscale suburb of Chevy Chase, Maryland, just north of DC. His new post required Lansdale to work extremely long hours, and he would occasionally disappear on assignment for days or weeks at a time, leaving Metta to raise their young girls largely by herself. "Mother will never forgive Franklin Roosevelt for taking her husband away," her daughter Chloe remembered. "She was angry about that all her life."

The Counterintelligence Group was primarily concerned with uncovering and eliminating spies or other internal threats to the military, orchestrated by foreign powers. Lansdale's first assignment under General Strong was reviewing and processing the many reports that came into Washington from the intelligence branches at military bases around the country.

Often these reports were concerned with identifying individuals within the military who were thought to pose a potential security threat. Lansdale quickly realized that the intelligence infrastructure was efficient at spotting these potential threats, but what to do next with these men and women once they had been flagged as suspicious had not yet been determined. When he pointed out this problem to his superiors, Lansdale suddenly found himself placed in charge of developing a solution. For his new task, Lansdale was first promoted to executive officer of the Investigation Branch and then eventually to chief of the Review Branch.

The threats that were of concern to Lansdale and the whole of the US intelligence community were varied, but chiefly stemmed from two sources: Nazi sympathizers and Soviet espionage. In the early 1930s, as the Third Reich amassed power in Germany, organizations of Americans, typically of German descent, who were sympathetic to the Nazis' ambitions began to form across the country. In 1936 the Amerikadeutscher Volksbund (or the German American Bund) was established in Buffalo, New York. Members of the Bund were openly supportive of the Nazi regime—they advocated for American isolationism and any other policy they felt would give Germany an advantage in the nearly inevitable coming war. Membership in the Bund was hard to estimate, but a "Pro-American Rally" held by the Bund at New York's Madison Square Garden in February 1939, which featured a towering portrait of George Washington flanked by swastikas on the main stage, drew a crowd of over twenty thousand.

Throughout the early years of the war, the Bund was active in circulating pro-Nazi propaganda, holding demonstrations, and even sponsoring an organization for children, modeled after the Hitler Youth. Outrage over their activities prompted an investigation in 1939 by the House Un-American Activities Committee, chaired by Rep. Martin Dies (D-TX), which found clear and direct ties between the Bund and the Nazi regime. When the United States entered the war against Germany in 1941, the Bund organization was officially outlawed, but this did little to eradicate the pro-Nazi sentiment throughout the country or to eliminate the threat of Nazi supporters interfering with the US military establishment.

Unlike Germany, which was clearly an American adversary, the Soviet Union was ostensibly an ally after the invasion of the Soviet Union by Nazi forces in June 1941. But the alliance was a tenuous and temporary one. It was widely accepted and understood that the USSR had cultivated an expansive network of spies and informants working in both the United States and the United Kingdom for the purpose of stealing wartime information. All those in the US intelligence network knew that while Russia may be an ally, it was certainly no friend.

It was in the context of investigating Soviet interference in the nuclear race that Lansdale first came into contact with the top-secret Manhattan Project. After a few months of working in the Counterintelligence Group, Lansdale had developed a reputation as a man who could get things done. So in February 1942, when James B. Conant, president of Harvard University and the chairman of the National Defense Research Committee, requested the army's help in addressing a security problem within the developing nuclear fission research project, Lansdale was the one who received the call to come to Conant's office.

It was in this meeting that Lansdale first learned about the possibilities of nuclear fission and the American effort to build an atomic bomb.

Conant had grown concerned about security at the Radiation Laboratory at the University of California, Berkeley. Headed by Ernest Lawrence, the scientists at Berkeley were working on designing and building what would be the world's largest cyclotron—an early particle accelerator that they intended to use in experiments for studying methods of separating the isotopes of uranium. Conant worried that the scientists were not taking the security risks to their research seriously, and their recklessness posed a threat to the project as a whole. Unsure what to do, he asked for Lansdale's help.

Lansdale thought the best way to assess the situation was to see what was going on for himself. Dressed as a civilian, he set off by train from DC's Union Station to San Francisco, arriving in California on February 19, 1942. He had hastily pieced together a cover story for his visit, hoping to dispel any suspicions about his sudden appearance at the university—he would say he was a lawyer living in Cleveland who had been born in the Bay Area and was planning to volunteer for military service, but he wanted to revisit his roots and complete some last-minute research before shipping out.

Early in the morning, the day after arriving at Berkeley, Lansdale set out on foot from the center of campus up a hill to the building where the cyclotron lab was being constructed. He assumed that at some point on his path toward the facility he was bound to be stopped by guards and turned away. But no barrier materialized. Instead, he passed through two unmanned and unlocked chain-link gates, flung wide open, hung with signs reading BLIND ROAD—NO VISITORS. Continuing his climb, Lansdale arrived at the new laboratory building, an enormous circular structure capped with a massive dome. From its vantage high atop the hill, the laboratory overlooked the campus below and the San Francisco Bay in the distance.

Lansdale tried the door of the lab and found it unlocked. He entered and spent almost two hours looking around at the construction, relatively unnoticed. Blueprints clearly labeled for the 184-inch cyclotron sat out on lab benches surrounding the central workspace, where an enormous mass of metal was being laboriously assembled into the particle accelerator. While the disk-shaped cyclotron itself would be large—over fifteen feet in diameter—it was dwarfed by the two immense circular magnets, weighing a total of 4,500 tons, that sandwiched the cyclotron from above and below. Supporting the massive weight of the magnets was a 3,700-ton stainless steel yoke, some thirty feet tall, that filled the room. The bottom half of this oblong support apparatus had been embedded into the foundation of the building during construction, and the domed structure had been erected around it.

Up to this point in the history of science, most of the instrumentation used in particle and nuclear physics had been small, homemade affairs, cobbled together on tabletops in dingy cramped labs by the scientists themselves. As the field developed and the need for larger, more powerful instruments increased, the construction of these colossal devices outgrew the capabilities of even the most ambitious graduate student. Lawrence's cyclotron was a colossal undertaking—the instrument itself cost almost $1.5 million. The industrial-sized instrument required industrial construction, and a crew of workers had to be brought into the laboratory, further expanding the circle of people who had knowledge of, and access to, the lab.

It was the construction workers, not the scientists, whom Lansdale first encountered. When the manager on duty finally approached Lansdale and asked who he was, Lansdale gave his alias and explained he had just been out for a walk on campus, had seen the new building, and had wondered what was inside. The manager was all too happy to show Lansdale around the laboratory, pointing out some of the more peculiar aspects of the room's construction and explaining that they were "trying to break the uranium atom" in pursuit of a "new explosive."

Descending the hill on his way back to the main campus, Lansdale, troubled by his experience in the lab, hoped that it was the construction crew and their loquacious manager who posed the greatest security risk. Surely the scientists, with a larger understanding of the cyclotron's place in the attempted development of new and unimaginably terrible weapons, would have more restraint. It was by then early afternoon, so, armed with a temporary pass and letter of introduction, Lansdale headed for the faculty club—the center of faculty life (and gossip) on campus—for lunch.

A popular location for professors to meet away from the bustling throngs of students, the faculty club's vaulted, wood-paneled main dining room, more in keeping with the decor of a Viking hall than a public university, was filled with round tables covered in white linen and set with china. One of these tables, set for ten and situated near the center of the room, was popularly referred to as the "physicists' table," as it was here that Professor Ernest Lawrence and his colleagues ate and socialized daily. Sitting at a table on the patio, with a view of this central spot, Lansdale watched that first afternoon as Lawrence and each of his fellow physicists arrived and took their seats.

Over the next several days Lansdale endeavored to gain introductions to as many of the men in this scientific clique as possible—a task that he found remarkably simple. Within a week, Lansdale had talked to nearly every member of "Lawrence's Brain Trust," save Lawrence himself.

More shocking than the willingness of these scientists to accept without question the sudden presence and manifold inquiries of a stranger with no apparent connection to nuclear research was the information they readily divulged. After just a few days of casual conversations, Lansdale learned nearly every detail of the scope and purpose of the cyclotron project. The scientists held little back, from the functioning principles of the instrument to its possible use in the separation of uranium isotopes, to the enormous potential explosive power contained in the fissile (^{235}U) isotope of uranium. Lansdale recalled that one scientist explained that "one pound of U-235 contained enough energy to lift all the buildings in San Francisco a mile high" and that "if the energy of fission could be suddenly released . . . it was plain that the discovery . . . would win the war."

Of all the men with whom he spoke, only the first, Donald Cooksey, showed any hesitation discussing the project. When Lansdale inquired about the cyclotron, Cooksey insisted that "any connection between the experiments with the cyclotron and the war was very far-fetched. . . . However they did have high hopes for its connection with the treatment of cancer."

Before returning to Washington to make his report to Conant, Lansdale went back to the cyclotron lab one more time, again encountering no barriers, to confirm for himself that he would have easily been able to walk away with instrumentation blueprints and other documentation that were left lying out.

Days later, Lansdale stood in Conant's office in Washington and gave a full report on the dismal state of security he had found at Berkeley. Horrified by what Lansdale had learned, Conant asked that Lansdale return to Berkeley, this time in military uniform, and address the security concerns he had seen firsthand.

A month after his initial visit, Lansdale made his dramatic return to the faculty club. The scientists were having lunch at their usual table when Lansdale, turned out in full, formal uniform, made his entrance. He strode across the room directly to their table with an air of authority that his civilian alter ego had lacked. The scientists, who were engaged in an energetic discussion about some confounding detail of the cyclotron's construction, did not see Lansdale until he reached their table. One by one the scientists fell silent, staring at the strangely familiar officer. Watching their faces, Lansdale tried to mask his amusement as each man began to put the pieces together, and his expression changed from that of a curious scientist to that of a child suddenly realizing he is about to be sent to his room.

Hoping that the shock value of his reappearance would make an impression, Lansdale admonished the scientists for their carelessness, reading aloud to them excerpts from his notes detailing exactly what information had been given to him and by whom, and attempting to "impress upon them how careless they were and how important it was to avoid telling strangers what they were doing." Most of the scientists seemed sufficiently chastened, with the exception of the reticent Cooksey, who proudly told Lansdale that he had been growing increasingly suspicious of his questions during his previous visit and had been about to report him to the FBI before he had left.

After handling the Berkeley matter for Conant, Lansdale returned to his usual job under General Strong until one day, in September 1942, as he was sorting through intelligence reports in his windowless office in the basement of the new Pentagon building, Brigadier General Leslie R. Groves appeared outside his door. Groves, a man whose massive stature matched his reputation at the Pentagon, had just been placed in charge of America's rapidly expanding nuclear program, and he wanted Lansdale's help with security. Groves was quick to tell Lansdale that he'd been selected not because of any superior quality he possessed but only because Conant had already let Lansdale in on the details of the secret project earlier that year, and Groves was endeavoring to minimize the number of people who knew about the project's existence. Lansdale accepted Grove's request. He received a promotion to lieutenant colonel, and at the age of thirty became the head of intelligence and security for the largest top-secret military operation in history.

Throughout his time leading the Manhattan Project, Groves earned an enduring reputation as a difficult man to get along with. He was abrupt and

General Groves awarding Colonel
Lansdale the Legion of Merit, 1945.
Public domain, courtesy of Chloe Pitard.

uncompromising, and often scathingly blunt in his assessments of others. While many of the men working under Groves found him incredibly challenging to work for, Lansdale's military education and upbringing by a similarly intractable father had prepared him well. Despite their bumpy start, the two men generally got along. For the rest of his life, Lansdale spoke very highly of Groves: "He was in fact the only person I have known who was every bit as good as he thought he was."

As with the more general army intelligence work, the ever-present threat of Soviet espionage in the Manhattan Project's research loomed large over Lansdale's new appointment. Much of the Soviets' efforts had become focused on obtaining information on the nuclear programs in both the United States and the United Kingdom. In many respects, these efforts were successful, despite Groves's obsession with project security, and the USSR was able to obtain a great deal of information about the American nuclear program—enough information, in fact, to enable Soviet scientists to build and test a nuclear weapon merely four years after the first American tests.

But Groves knew his nuclear program faced an even bigger threat: the German nuclear program had a two-year head start on the American effort, and while there had been no indication that the German project had yet produced any kind of weapon, Groves and others feared that whispered reports of an accelerating American nuclear program would spur Germany to accelerate their research in kind. Groves was certain that the ultimate success of the Manhattan Project lay in maintaining absolute secrecy around what the American scientists were doing. This secrecy became Lansdale's charge.

Working under Groves, Lansdale established a special Counterintelligence Corps for the Manhattan Project that initially consisted of 25 officers and 137 enlisted men. The officers recruited for this top-secret group were mostly borrowed from each of the intelligence commands operating in different regions around the country and reassigned to the Manhattan Engineer District, the code name for the administrative arm of the Manhattan Project. Lansdale's shadow organization of "creeps" operated outside the normal intelligence channels, keeping their own records and reporting to Lansdale directly. The Manhattan Engineer District Counterintelligence Corps grew alongside the project itself, as the demands of the ever-growing number of sites increased the group's number to several hundred men.

The Counterintelligence Corps was responsible for most of the project's security concerns, including background checks on recruited scientists and securing the airspace above the main Manhattan Project sites of Oak Ridge, Hanford, and Los Alamos. During one memorable incident, Lansdale received a late-night phone call about an unidentified aircraft that had been spotted flying over the Hanford site. Recent reports of Japanese submarines operating off the coast of Washington State had put everyone on edge, so a pursuit squadron from the Western Defense Command was sent to investigate. The aircraft turned out to be just a private plane belonging to a very startled civilian who had gotten lost in the fog after failing to file a flight plan.

Beyond the day-to-day security needs of the project, Lansdale's role as the head of such an extensive top-secret operation came with its own unique set of challenges when well-meaning individuals who were not among the chosen few occasionally stumbled a little too close to the truth for comfort. When DuPont, a chemical manufacturer and a major corporate contractor working within the Manhattan Project, was targeted for antitrust violations by the Department of Justice, which remained unaware of the role the company was playing, Lansdale called Assistant Attorney General Thomas Clark to persuade him to drop the matter—no questions asked. In another instance, Groves was horrified to discover that President Roosevelt had been sending copies of classified documents about the project to Supreme Court associate justice Felix Frankfurter. Lansdale contacted Frankfurter, explained the need for utmost secrecy surrounding the project, and asked him to return all copies of the documents that he had received, which he kindly agreed to do.

Lansdale's team was also responsible for sorting through and making sense of what little intelligence was coming out of Germany regarding their own nuclear program. As the Manhattan Project was careening toward a functioning atomic weapon, American scientists were certain that their German counterparts were working at a similar breakneck pace. But most of the information that was making its way from Germany told a very different story. Letters from German scientists and other informants on the Continent indicated a modest, academic-scale nuclear program. Copies of a handful of scientific journal articles published by German physicists obtained from Switzerland discussed the results of preliminary experiments, making no indication of a critical reactor pile.

But Groves and his team remained doubtful of these reports; surely, the Germans were trying to disguise the true scope of their project with red herrings. That the renowned German physicists could be falling so far behind was simply unfathomable. As the Allies prepared to invade Italy in 1943, Lansdale saw an opportunity to send a specialized covert scientific intelligence mission into Europe. The team, which would comprise both military personnel and scientists, would be tasked with capturing German scientists, documentation, and equipment while trying to determine the true extent of Germany's atomic program.

Approval for a project of this scope required sign-off from not just General Groves and the chief of staff but also from each division of the War Department general staff. After writing up the plan for the mission in a memorandum, Lansdale personally carried a piece of paper to each of the divisions at the Pentagon for their signatures. He was surprised to find that the idea for some sort of mission to enter Germany with the express purpose of capturing scientific information had occurred to almost everyone with whom Lansdale spoke. But since nothing like this had ever been tried before, few divisions had enough clout to implement such a project. The Manhattan Project, with its high-priority status and specific and pressing objective, allowed Lansdale to do what others could not. Within days the Alsos Mission had received authorization.*

---

* *Alsos* is Greek for "grove," a nod to General Groves. When Groves learned of the significance behind the code name, he was furious, but reasoned that changing it would be even more likely to draw unwanted attention.

5 | THE SOLDIER: BORIS PASH

THE UNORTHODOX ALSOS MISSION would require a special type of commander. Someone experienced in covert operations and intelligence gathering, and who could think quickly on their feet. Someone who understood the complex politics of the military machine, but who would not allow bureaucratic red tape to hinder their work. In late 1943, as the plans for the Alsos Mission were beginning to take shape, Groves and Lansdale identified Lieutenant Colonel Boris T. Pash as the right person for this job.

Boris Pash was born Boris Fyodorovich Pashkovsky at the turn of the twentieth century in San Francisco. His father, Fyodor Pashkovsky, was a prominent member of the Russian Orthodox clergy who had been sent to the United States by the church. There, he had met and married Boris's mother, Ella Dabovich, sister to a well-known Serbian American cleric. Years later, Boris's father would go on to become the Most Reverend Metropolitan Theophilus, the highest-ranking cleric of the Russian Orthodox Church in North America.

Despite being born into a priestly family, Boris was destined for the battlefield, not the pulpit. One of those rare, enigmatic figures who seem fated to find their way into the history books, Pash would find himself inserted into many of the greatest moments of the twentieth century.*

---

* Pash has, in more recent years, found his way into the digital age as well. His potential involvement in the Kennedy assassination is a popular topic of debate in some of the dark and delusional corners of the Internet. A fictionalized version of Pash is also a character in a popular video game franchise.

When Pash was twelve, his father was recalled to Russia and moved his family back to his home country. When World War I broke out, young Boris served as a private in the Fifty-Second Artillery Division of the Russian Army alongside his father, who worked as a military chaplain. Soon after armistice, the Bolshevik Revolution swept Russia, and Pash continued his military career, serving in the anti-Communist White Navy as an interpreter for British forces operating in the Black Sea.

It was during this time that Boris was first introduced to Lydia Vladimirovna Ivanova. The young blonde aristocrat, who, even in the waning days of the Russian aristocracy, ranked far above Boris's station, initially paid the young American no mind. But Boris, already utterly unconcerned with the accepted rules standing in his way in pursuit of a goal, was undeterred. Learning that the object of his affection adored cats, Pash set out into the Ukrainian countryside early one morning. He returned with a small, fluffy white kitten, which he had named Puss, who served as his introduction to Lydia. The couple were soon married, and Lydia would take on the feline's moniker as an affectionate nickname used in letters and telegrams throughout their marriage.

When Lenin and his Communist Party took control of Russia, Pash and his new wife departed the country, making their way first to Berlin, where their son Edgar was born, and then to Springfield, Massachusetts, where Pash studied for his bachelor's degree in physical education. The small family eventually made their way back to Pash's native California, where Pash took a position teaching PE at Hollywood High School in Los Angeles.

Unwilling, or perhaps unable, to leave military life behind entirely, Pash joined the US Army Reserve in 1930, and for the next decade was called up occasionally for brief stints of duty with the Military Intelligence Division. In 1940, as the US military began to mobilize for war, Pash was officially called into active duty, serving with the newly formed Western Defense Command (WDC), a branch of the Military Intelligence Division, operating out of the Presidio of San Francisco, a historic military fort at the foot of the Golden Gate Bridge. At thirty-nine, Pash was already considerably older than most of his fellow officers. His son Edgar, who was old enough to serve as well, joined his father at the WDC in the Naval District Intelligence Office.

As with most of the intelligence gathering centers, a primary concern for those working at the Western Defense Command was warding against Soviet espionage and interference. The presence of the Russian consulate in

San Francisco, as well as the reputation of Berkeley as an enclave of radical liberalism, made the Bay Area a primary target for these anxieties. Pash's Russian background made him uniquely qualified, and he was soon responsible for tracking and assessing Soviet threats on the West Coast as part of the covert security network for the Manhattan Project under Colonel John Lansdale.

Pash's experiences during the Russian Revolution had cemented an extreme disdain for the Soviet Communists, a view he carried over into his investigations, which he executed with a brazen zealotry. Lansdale recalled that "a number of very intense investigations of the activities of several Communist Party members who were on the [Manhattan] project were conducted by Pash. These people were followed, and clandestine microphones were installed in their houses and in places they frequented."

The subject of Pash's most intense investigation was none other than J. Robert Oppenheimer, one of the leaders of the scientific efforts of the Manhattan Project, who was working at Berkeley in the early years of the war. While never directly a member of the Communist Party himself, the reputations of Oppenheimer's wife, Kitty, as well as some of their friends, coupled with his own liberal political leanings, had raised several red flags, prompting an FBI investigation of the scientist. Pash's team was also asked to investigate.

In the late summer of 1943, Pash conducted an interview with Oppenheimer, which, like many of his other investigations, he secretly recorded. These recordings would serve as the basis for the grueling security hearings Oppenheimer was subjected to following the war, in which his nonanswers to certain questions were taken as proof of Communist sympathies.

The questionable loyalties of his lead scientist saddled Groves with the difficult decision of how best to proceed. As Lansdale put it, "There was plenty of 'smoke' and some amount of fire" in the Oppenheimer dossiers, and under ordinary circumstances, a man with Oppenheimer's records would not have been cleared for work on the Manhattan Project. But after "a rather aggressive search for someone else who had the combination of qualities necessary for this job," Groves determined that Oppenheimer was simply the only one who could do what needed to be done and moved forward with the necessary clearance.

Pash, however, remained staunchly opposed to the scientist's involvement in the project, and it was later suggested that Groves may have been motivated, at least in part, in his selection of Pash to lead the Alsos Mission in order to get Pash out of California and away from any further involvement with Oppenheimer.

In April 1944 Pash officially took command of the Alsos Mission. The planned covert intelligence unit would be small, composed of no more than a couple of dozen men. The investigations that the mission was tasked with carrying out would be conducted by civilian and military scientists, who would be supported in their movements and work by a small number of infantry, interpreters, and Counterintelligence Corps operatives. In order to enable their easy movement from location to location, Alsos would not be formally attached to any other unit but would instead operate independently under Pash's command. The thin, bespectacled lieutenant colonel, twice their age, with thinning hair, stood out among the gaggle of young men he led, who referred to their commander affectionately as "Chief" or "Old Man."

The task ahead of Pash and his team was unique and difficult. They had to search for and destroy a superweapon that no one was certain even existed: "We were going to risk our necks in the hope of finding nothing!"

It was imperative that the Alsos Mission determine the true extent of the German program as quickly as possible, but it was also equally important that they be the *first* to find and capture all supplies, equipment, and scientists associated with the German nuclear program before the advancing Soviet forces had a chance to interfere—an aspect of his charge that Pash no doubt relished.

Colonel Boris T. Pash, drawn by Michael Thurgood. *Emilio Segrè Visual Archives, Gift of Michaele Thurgood Haynes and Terry Thurgood, Thurgood Collection*

6 | ALSOS IN ITALY

Italian Peninsula
Summer 1943–Spring 1944

IN JULY 1943 ALLIED FORCES under British Field Marshal Bernard Montgomery and US General George Patton landed on the island of Sicily, just south of the Italian mainland peninsula, giving the newly formed Alsos Mission its first opportunity to access Europe.

No one expected that a mission to Italy would turn up much information about the German nuclear program directly. Intelligence gathered from sources in Europe suggested that there was not much overlap between German and Italian scientists, and it was doubtful that the Nazis would allow such a high-profile project as nuclear weapons development to be conducted beyond German borders. But as Major General Strong wrote in a memo to General Marshal outlining the plans for the mission, "While the major portion of the enemy's secret scientific developments is being conducted in Germany, it is very likely that such valuable information can be obtained thereon by interviewing prominent Italian scientists in Italy." It was possible that the Italian scientists, who had been mostly able to maintain contact with their friends and colleagues in Germany, could, through correspondence or gossip, have some clues as to what type of work the Germans might be doing.

And while they had likely not been working on nuclear research, the Italian scientists had certainly not been idle. To help disguise the true focus on nuclear research, the official scope of the Alsos Mission extended to any and all scientific material that might assist in the Allied war effort. By the end of the war, Alsos would generate hundreds of reports containing valuable scientific intelligence in nearly every scientific discipline.

Sending the mission into Italy had one additional benefit. Nothing quite like the Alsos Mission had ever been attempted. Italy provided the unusual outfit with, as Lansdale put it, the "unique opportunity . . . [of] a dry run or exercise to prepare it for the effort to acquire information about the German atomic program after the Allied landings in Europe."

On Saturday, November 20, 1943, Boris Pash was at home in the Bay Area of Northern California, making plans with Lydia for a Thanksgiving dinner with friends later that week, when the phone rang. It was the call he had been waiting for since he had begun working on plans for the Alsos Mission in October: General Strong's office was ordering Pash to report Washington immediately. A few days and a missed turkey dinner later, Pash found himself in the office of the US secretary of war, Henry Stimson. The Alsos Mission was finally being mobilized, Pash was told. He and his men were headed to Italy.

During their meeting, Stimson provided Pash with two letters. One was an introduction to General Dwight D. Eisenhower, who was commanding Allied forces in Italy and Northern Africa. The other was to General Walter Smith, Eisenhower's chief of staff. These letters indicated that Pash's task was of the utmost importance, and that his team was to be assisted in any way that they might require. This request for support without any kind of explanation was an unusual break in typical protocol that signaled to Pash the urgency felt by General Groves and General Strong about the necessity and importance of the task he was about to undertake. Similar letters would be provided months later during Alsos's time in France and Germany, allowing Pash and his team to move quickly and freely around the front, borrowing jeeps, armored cars, and enlisted men on an as-needed basis.

Pash was next introduced to the men he would be leading. By design, the team was small. James Fisk of the Bell Telephone Company would lead the scientific operations, along with John Johnson from Cornell University, Lieutenant Commander Bruce Old from the Office of Naval Research and Development, and Major William Allis, from the War Department scientific staff. In addition to the scientists, the group would also include four Counterintelligence

Corps (CIC) agents who would join the group in Europe, four interpreters, and a small infantry unit. Of this group, only Pash and Fisk were fully briefed on the true purpose of the mission and their focus on atomic research.

Pash and the scientists began their journey with a flight from Washington, DC, to Natal Air Force Base in Brazil, before crossing the Atlantic and landing in Dakar, Senegal. The group then had to move northward, to Marrakesh, Morocco. A brief stopover at a small emergency air base in Atar, Mauritania, resulted in a blown tire, and the group was forced to spend the night in this remote Saharan town while a replacement was located. A storm further delayed the team, and they finally arrived in Algiers, Algeria, on December 14, nearly four days later than expected.

Upon his arrival, Pash was immediately ushered into general headquarters, where he handed the introductory letters that he had been given in Washington to General Smith.

The letters did their job, and the next day, December 15, the Alsos team boarded a flight to mainland Europe.

In Southern Italy, the Alsos Mission established their first headquarters, taking over the director's suite at the Bank of Naples. Pash and his scientists met up with the rest of the new Alsos team a few days later, at a Christmas party thrown by the CIC detachment in the city. Ralph Cerame from Rochester, New York, had been selected for his Italian language skills. Carl Fiebig, a former furniture salesman from Sebewaing, Michigan, had grown up in a largely German-speaking community. His ability to speak German without any hint of an American accent had earned him a spot on the Alsos Mission as well. Perry Bailey had no special language skills, but his energy and enthusiasm stood out.

For their fourth team member, Fiebig pulled Pash aside at the party to make a suggestion: "Colonel," Fiebig said, "my buddy, Gerry Beatson, from Rockford, Illinois, is a fighting fool of a CIC agent. I know he'd like to join us."

Not by accident, Beatson had come up to Naples from his post at Sorrento and was in attendance at the party, waiting anxiously in the corner for a signal from his friend to come over and meet Pash. Fiebig waved him over and Pash immediately liked the look of the young man, who had previously worked in home finance but had become quite an accomplished fighter and marksman.

"Well Gerry," Pash said to a beaming Beatson, "we expect to operate right up with the infantry and will need men with your experience. You could be my trigger man."

By the time Alsos arrived in late 1943, the American and British forces had been in mainland Italy for several months. Italian dictator Benito Mussolini had been deposed and imprisoned earlier that fall, and Germany's Italian allies had already surrendered in September. The Nazis occupying Italy, however, had not. While the Allies had managed to gain a foothold in the southern part of the country, the Germans were making effective use of the mountains and rivers in central Italy, forming a strong defensive line across the country from the Tyrrhenian to the Adriatic Seas. Alsos's two primary targets in the country, Edoardo Amaldi and Gian Carlo Wick, nuclear physicists whose names had been provided by Enrico Fermi himself, were both located in Rome. But as long as Allied troops remained stalled in their progress north, Alsos's investigations were limited to the cities and towns along Italy's southernmost coast.

During their first week in Italy, Pash worked to establish understandings with the Allied commanders as well as the Italian provisional government about what his mission was hoping to uncover. By Christmas, Pash reported to Lansdale back in Washington that Alsos had "become substantially independent of any formal organizations in this theater," meaning they were able to move freely around Allied territory as the need arose.

As Alsos agents and scientists set to work tracking down and surveilling any targets of interest in Naples and throughout Southern Italy, they quickly discovered that there was not much work remaining to be done. Many, if not most, of the scientists and laboratories in Southern Italy had already been handed over to the Office of Strategic Services (OSS), the primary US intelligence agency during the war, which had been in the area for three months before Alsos's arrival. In many cases, Alsos was left to review information that had already been collected, and to reinterrogate scientists who had already shared much of what they knew.

Within just a few weeks, all leads that could be followed from their post in Naples had been scrutinized, and the mission was largely left with nothing to do. If Alsos was to make any significant progress in Italy, they would need to find a way to access and interrogate scientists working in the capital.

On January 21, 1944, Pash was informed of the plan for an imminent large-scale landing at Anzio, on the western coast of Italy, just south of Rome, as the Allies tried to circumvent the Nazi blockade. It was decided that Alsos would travel to this new front and join with the S-Force, a conglomeration of several small intelligence-focused units including American, British, French, and Italian military and technical personnel, which would enter Rome at the first opportunity in order to secure their targets and protect them from damage.

But like so many of the Allied offensive attempts in Italy, the Anzio operation did not go as anticipated. While the Allies were able to establish control of a portion of beachfront, their progress inland stalled again almost immediately. As the days wore on, the hoped-for movement toward Rome never materialized. On January 31, Fisk and Johnson returned to the United States, leaving Pash with a depleted technical staff.

In a last-ditch attempt to continue his investigations in Italy, Pash decided to try a different approach. If Alsos was unable to access and interrogate their targets in Rome, they would instead attempt to covertly extract the wanted scientists from the city and bring them back to their headquarters in Naples. A covert operation of this type would require the assistance of the OSS, and Edoardo Amaldi was selected as the target. Amaldi had studied under Enrico Fermi before the war, and it was hoped that he, if anyone in Italy, might have some insight into the German's nuclear research program. The plan, code-named Operation Shark, was to drop an undercover agent into Rome who would then locate Amaldi. Assuming that the scientist was amenable to early extraction, given the likely fighting in Rome in the months to come, Amaldi would be whisked off to an Adriatic beach.

A few weeks later, Pash was contacted by the OSS and told that the plans had to be changed. Amaldi would now be picked up from a western beachfront, which would require the use of two navy patrol boats. This presented a problem, as the navy had recently barred OSS agents from accessing these vessels. Some agents had apparently made a habit of taking high-ranking visitors on tours of the coastline and would, on occasion, direct these boats to travel within range of the German costal guns, in hopes that the enemy's gunners on the shore would send a few shells whizzing over their heads, giving their visitors a dramatic tale of enemy fire to tell upon their return to Washington. On one of these trips, the gunners had actually hit their mark, shooting up a couple of patrol boats and injuring some crew. If Pash wanted the navy to lend their craft to the OSS for Operation Shark, he would have to go to the top of the ladder. Pash boarded a transport back to Algiers to request help directly from Admiral Lewis, who grudgingly agreed to provide the necessary boats.

Pash returned to Naples just in time for a sudden transport up to the beachfront at Anzio. An unexploded glider bomb had been discovered on the beach and was being protected by Allied forces. On February 9, Pash, Major Allis, and a few other Alsos men landed on the beach, ducking under German fire. They slowly made their way to a command post that had been set up in the basement of an abandoned but (to the delight of Pash's men) still-stocked wine cellar. The next day they made their way to the crater in the sand that held the unexploded bomb. While not related to the nuclear program, the unexploded weapon provided Alsos with their first opportunity to examine advanced Nazi technology. The team worked quickly, crouched behind empty gasoline drums with sand and rocks flying around them as German planes dove down from above firing into the ground. The reward for their persistence was a full set of images and specs on the bomb.

Operation Shark did not yield the same kind of positive results. Pash was not given much information by his OSS contacts. They had found Amaldi, Pash was told, but a first attempt at extraction had failed and a second attempt had aroused German suspicion, resulting in increased patrols. Further attempts would be too risky, and Operation Shark was suspended indefinitely. Pash's last remaining hope of Alsos interrogating any prominent Italian scientists before the capture of Rome faded.

Several months later, in June 1944, Pash would finally get his man. When interrogated by Pash, who quickly made his way back to Italy as soon as Rome was liberated, Amaldi insisted that he had never been contacted by any OSS agents as Pash had been told. Pash was so incensed by this apparent breach of trust that he vowed to never involve the OSS in Alsos business again. He was left to wonder how exactly the boats he had acquired permission for had really been used.

By March 1944 Pash and most of the remaining Alsos members had returned to Washington, leaving agents Cerame and Bailey to man the office in Naples. As had been expected by Groves and Lansdale, the Alsos Mission to Italy had not revealed any significant intelligence on the German atomic program. Still, as Pash put it, "Sound intelligence is seldom served up on a platter. It must be developed bit by bit," and Alsos's time in Italy had resulted in a few significant wins for American scientists interested in other areas of

Nazi wartime research. The mission scientists were able to submit reports on a variety of topics, including radar, guided missiles, and chemical warfare. Vannevar Bush wrote to Groves that "the information obtained by the Mission which relates to the work of the NDRC, has been most significant and one or two items have, in my opinion, justified the whole enterprise."

After their return to the United States, Groves made the decision to continue the Alsos Mission on a permanent basis, with Pash as the commanding officer. This next iteration of Alsos would soon be making its way to France and Germany with a few changes in mission operations. Their time in Italy had illustrated that in order to be effective, Alsos needed to be as close as possible to the front, ready to move in as soon as cities and towns were secured by the Allied troops. Waiting in the wings had left the mission too isolated from the front and at the mercy of information provided by others who might not appreciate the true urgency of their task. It also left too much time before their arrival for sites to be looted, or even accessed by intelligence agencies from other countries. If Alsos wanted to be the first and only unit to obtain information on the German nuclear program, they needed to be the first ones on the scene at every one of their targets.

Another change involved the scientific staff. Instead of a group of scientists all working under Pash's direction, Alsos's command would now be split in two. The scientists would be directed by a scientific chief who would be responsible for identifying targets for the team based on the movements of the front and developments in the information that they gathered along the way. Pash would remain the military chief of Alsos, and he and his men would be responsible for going where the scientist pointed them and securing those sites for investigation.

All that remained was to choose a scientific chief. Experience in Italy had taught them that not all the information that they would encounter would be reports and other summary documents that are easy enough to interpret. Most scientific intelligence, they now understood, would be in the form of raw data, laboratory notebooks, and detailed high-level conversations with captured scientists. Interpreting this type of scientific information would require someone deeply familiar with the subject matter. They would need more than just a scientist for their mission into Germany; they needed a world-class nuclear physicist.

7 | THE SCIENTIST: SAMUEL GOUDSMIT

BORN IN 1902 IN THE HAGUE to a blue-collar Dutch-Jewish family, the notion that Samuel Goudsmit would become a scientist—much less a world-renowned physicist whose career would include forty-eight Nobel Prize nominations—seemed unlikely. Describing his family, Goudsmit once summarized: "My grandfather . . . was a tourist guide in Hotel des Indes in The Hague, my mother had a millinery, and my father a wholesale business in seats . . . for water closets." As for young Sam, he wanted to become a detective. As a child, the eager boy would spend hours every week camped outside the local police precinct, watching the comings and goings of the officers and reading detective novels.

But by high school, the labyrinthine puzzles of science had captivated Goudsmit's attention, leading him to enroll in the physics program at Leiden University. It was there, in the early months of his studies, that Goudsmit met physics professor Paul Ehrenfest. Charmed by the boy's earnest eagerness, Ehrenfest immediately took an interest in young Goudsmit's progress. It was Ehrenfest who made the fateful suggestion to Goudsmit that, while accompanying his father on a trip to Germany, he might stop by the lab of Friederich Paschen, a spectroscopist working in the southwestern town of Tübingen. Paschen, like many physicists working in the early twentieth century, was attempting to decipher the internal structure of the atom and was doing so by studying the discrete wavelengths of light emitted from the electron orbitals of different ionized gases. Goudsmit was captivated by these experiments; the spectral lines were like a shining trail of clues to nature's invisible secrets just waiting for someone to unravel them. Goudsmit returned to Leiden with a plan to study these spectral lines himself.

Goudsmit began working, again at Ehrenfest's insistence, with fellow physics student George Uhlenbeck. This collaboration proved productive. The pair
soon noticed a peculiar anomaly in the lines of light produced when their
experiment was placed in a magnetic field: some of the lines seemed to split in
two. Goudsmit and Uhlenbeck posited that these anomalies could be explained
if all electrons possessed an intrinsic "spin," either clockwise or anticlockwise,
as they flew around their atomic orbitals. The idea of electron spin had been
considered and rejected by many more-seasoned physicists who noted some
larger conceptual issues that this theory introduced. But young Goudsmit and
Uhlenbeck, undaunted by these stumbling blocks, forged ahead, publishing a
high-profile article in the journal *Naturwissenschaften* in 1925. Their discovery,
which put into place a critical missing piece of the subatomic puzzle, gained
Goudsmit and Uhlenbeck a level of international scientific acclaim.

As Goudsmit and Uhlenbeck were completing their graduate work in the
spring of 1926, Walter Colby, a physics professor at the University of Michigan,
arrived in Leiden. Colby was attempting to create a theoretical physics group
within his department and was in Europe looking for fresh talent in the field.
Ehrenfest immediately suggested to Colby that he hire both his young protégés.

Goudsmit was in Germany visiting Paschen once again when he received
Colby's offer of a professorship in Ann Arbor. He was initially hesitant to
accept, especially before he had had a chance to discuss the matter with any
of his colleagues in the Netherlands or with his new wife, Jaantje Logher. But
unbeknownst to Goudsmit, Ehrenfest had already arranged the whole matter,
once again shepherding Goudsmit toward a fateful next step. A few months
after graduation in July 1927, Sam Goudsmit and Jaantje along with George
Uhlenbeck and his wife, Else, found themselves aboard the RMS *Baltic* on the
nine-day voyage from Liverpool to New York City.

When they arrived in New York and disembarked from the ship, standing
on the pier waiting to collect them was none other than J. Robert Oppenheimer,
one of the future scientific leaders of America's nuclear program. At only
twenty-three, Oppenheimer had, like the new Dutch arrivals, only just completed his doctorate that past spring. During his graduate work he had spent
some time in Europe, which included a stay in Leiden during which he had
met and befriended Goudsmit and Uhlenbeck. For a few days, Oppenheimer
played tour guide for his Dutch colleagues, showing them the sites around New
York before the foursome continued their journey west by rail.

The arrival of the two new professors in the sleepy university town of Ann Arbor, "both looking thin, energetic and incredibly young for such well-known figures," blew new life into the small physics department. As originators of the theoretical physics curriculum at Michigan, Goudsmit and Uhlenbeck established the basic coursework requirements. They also established a weekly seminar series and journal review club, both of which had been prominent aspects of their own studies in Leiden under Ehrenfest.

Goudsmit struggled at first with his new life in America. The stark surroundings of the midwestern American city lacked the vibrancy, beauty, and sophistication of the European centers in which he had been raised. Coupled with the dreary weather and isolation from so many of his family and old friends, Goudsmit was continually considering alternative teaching posts in Europe.

Slowly, however, life in Ann Arbor began to improve. In 1929 Goudsmit was offered a teaching post in Zurich, and the University of Michigan, in an effort to convince him to stay, offered a sizable raise and assured him that he would no longer have to teach first-year classes, which he had hated almost as much as his befuddled students did.

Goudsmit was also able to find time during these early years in Michigan to continue the pursuit of one of his other lifelong passions: Egyptology. Goudsmit had become enamored with the study of the ancient culture during his years at Leiden University, when he had been encouraged to find a topic beyond atomic structure to present at a student discussion group. Interested by what he learned, Goudsmit enrolled in an Egyptology course, in which he was the only student that semester. Soon, Goudsmit had learned to decipher hieroglyphics and had begun collecting small Egyptian artifacts. He continued his studies and his collection expanded while he was at the University of Michigan. The Samuel A. Goudsmit Collection of Egyptian Antiquities is now a centerpiece in Michigan's Kelsey Museum of Archeology.

But perhaps the brightest spots of Ann Arbor life for Goudsmit were the annual summer physics gatherings at the university, which drew some of the best minds working in Europe and the United States to small-town America for a few weeks of lectures and discussions. Among the frequent participants were Niels Bohr, Enrico Fermi, and Goudsmit's old friend and trailblazer in the early days of quantum mechanics, Werner Heisenberg, who often stayed with the Goudsmits during these trips.

Left to right: Samuel Goudsmit, Clarence Yoakum, Werner Heisenberg, Enrico Fermi, and Edward Kraus. Ann Arbor, Michigan, 1939. *AIP Emilio Segrè Visual Archives, Crane-Randall Collection*

In 1939 Goudsmit was once again offered a post in Europe, this time at the University of Amsterdam. Turning down a post that would take him and his family back home to the Netherlands to be with his ailing parents was difficult. But Goudsmit was well aware that many of the best physicists in Europe, particularly those of Jewish descent, had for several months been steadily leaving the Continent for posts in America. Goudsmit began to regularly receive letters from European colleagues asking for his assistance in securing work in the States. Fearing for the safety of his family should he move them back to Europe, Goudsmit reluctantly declined the Amsterdam offer. He would later write in a letter to his daughter, Esther, that he felt sure that had he taken this post, they would not have survived the war.

For a time, Goudsmit's work continued at Michigan largely unaffected by the war in Europe, until early 1941, when Goudsmit received a message from Edwin Kemble, a friend and physics professor at Harvard University, asking if he would be able to come out to Cambridge for a few months to teach. A sudden shift to war-related research projects was gutting the halls of many universities as professors and researchers were called into war work, and Harvard had been no exception.

Over the previous few months, Goudsmit had noticed a slow but steady change in his academic circles. Friends and colleagues who had once openly and endlessly discussed their research at dinners and parties were suddenly falling silent when the subject was broached. Soon, these same scientists began disappearing altogether, vanishing without a word or a trace from their class-rooms and laboratories.

Where they were going and what they were working on was both the best and worst kept secret in science. The uranium problem had been loom-ing large since Meitner and Frisch's publication in January 1939. As more and more physicists, chemists, and engineers disappeared, those left behind readily surmised what they were up to: they had gone to work on a nuclear weapons program.

Despite his prominent work in the field of nuclear physics, Goudsmit was never included in the ranks of these disappearing bomb builders. Though he was never sure of the reasons behind this decision, he suspected that it had to do with his mother and father, both of whom were still in occupied Hol-land. Goudsmit assumed that those in charge of the nuclear program's security worried that Goudsmit's concern for his parents could make him vulnerable to sharing secret information in exchange for their safety.

But the Manhattan Project was not the only wartime research that required attention. Though not as clandestine as the nuclear work, developments in many fields including aeronautics and radar were also considered high-priority projects. Goudsmit was eager to do his part, and so, after accepting Kemble's invitation to teach at Harvard, he also began work on radar research across the Charles River at the Massachusetts Institute of Technology.

It was while at MIT in early 1944 that Goudsmit was approached by Vanne-var Bush, the head of the Office of Scientific Research and Development, about joining another top-secret project—an expedition into Germany to determine whether the Nazis had been successful in developing an atomic weapon. Goud-smit was not at all surprised that such a mission was being assembled. He had, in fact, written a letter to the head of MIT's Radiation Laboratory suggesting just this, and offering himself as a participant. But when the opportunity actu-ally presented itself, Goudsmit had mixed feelings. Conducting a high-level investigation was a childhood dream come true, and he relished the chance to play an important role in the war effort. But he worried about leaving his family and research behind. He was also quite concerned about actually experiencing life near a war front, and the anticipated lack of creature comforts he knew he was likely to encounter on the Continent.

Goudsmit was not alone in these concerns. Boris Pash was equally skeptical that a true academic scientist like Goudsmit would be able to cope with the rough conditions of the battlefield. Recalling his first encounter with the scientist, Pash wrote, "He told me, as we shook hands, that he liked the comforts of the civilized world and the quiet of a peaceful laboratory. Braving enemy fire or parachuting behind enemy lines was not his idea of recreation," which, from Pash's perspective, was not a good sign.

But as far as both Bush and Groves were concerned, Goudsmit was the best possible choice for scientific chief as Alsos prepared for their mission ahead into France and Germany. He was a talented and highly competent nuclear physicist and he was fluent in several of the major European languages, which meant that he would be able to quickly understand and interpret the significance of any data or equipment that Alsos might come across during their investigations. His close relationships with many in the European science community increased the likelihood that these individuals, once captured, would be willing to share what they knew with him. And finally, Goudsmit's lack of any formal knowledge surrounding the work on the Manhattan Project meant that he would not pose a serious security threat in the event of his capture.

The matter was soon settled and Goudsmit was brought to Washington to be briefed on the Alsos Mission, meet his fellow team members, and begin planning. Though their initial meeting was awkward, Pash and Goudsmit were quickly able to find their footing as Goudsmit's morose sense of humor was met with Pash's own dry wit. "I promised that if any jumping were involved," Pash assured the scientist about the days ahead, "I would go first and prepare a soft landing for him."

In reality, despite all the fussing and worrying, Goudsmit needed little motivation to jump into the war. For the Dutch Jew, the war against the Nazis had become personal. When Germany invaded Poland, Goudsmit had begun to take steps to get his mother and father to safety in the United States. The process took many months, but eventually Goudsmit's parents were issued American visas. But by the time the paperwork came it was already too late. The hard-won travel papers arrived at his parents' house just four days before the Nazi invasion of the Netherlands, and their door to escape was suddenly slammed shut. Goudsmit would receive one final letter in March 1943 from his parents, with the Auschwitz concentration camp listed as the return address. He would never hear from them again.

In the days after Germany's surrender in 1945, Goudsmit was able to unearth the records of what happened to his parents:

> The world has always admired the Germans so much for their orderliness. They are so systematic; they have such a sense of correctness. That is why they kept such precise records of their evil deeds, which we later found in their proper files in Germany. And that is why I know the precise date my father and my blind mother were put to death in the gas chamber. It was my father's seventieth birthday.

In Washington, in the days before the Allied invasion and before Alsos moved into Europe, Goudsmit threw himself into planning the Alsos Mission. As far as he was concerned, there was only one man who could possibly be leading a major nuclear research program in Germany: his old friend and classmate, Werner Heisenberg. Finding Heisenberg and his lab became Alsos's top priority.

8 | ALSOS IN ENGLAND

United Kingdom
Fall 1943–Summer 1944

THE UNITED STATES WAS not alone among the Allied nations in the pursuit of nuclear power. Great Britain and Canada had jointly begun their own nuclear research program, code-named Tube Alloys, shortly following the publication of the discovery of fission. But despite Tube Alloys' early start, it never reached the same enormous scope and scale of the Manhattan Project. In August 1943, in what would come to be known as the Quebec Agreement, Prime Minister Winston Churchill and President Franklin Roosevelt signed papers merging the two efforts.

With the Manhattan Project and Tube Alloys working together, General Groves and Colonel Lansdale hoped to take advantage of, and perhaps gain some control over, Britain's intelligence-gathering network. While Pash and the first iteration of the Alsos team had been making the best of their time and efforts in Italy, General Groves had dispatched Colonel Lansdale along with his chief of foreign intelligence, Major Robert Furman, to London in late 1943 to establish an Alsos office across the Atlantic.

Unlike the intelligence network in the United States, where the concept of scientific intelligence gathering was relatively new, the British Secret Intelligence Service (SIS) had been collecting information on prominent German scientists since the very beginning of the war. By the time the Alsos Mission began its Italian operations, British intelligence officials were confident in their belief that the German nuclear program was not going to pose a serious threat to the Allied war effort.

Their evidence for this belief was largely circumstantial. British intelligence agents had contacted Lise Meitner, both because of her study of fission and her close relationship with the German chemist Otto Hahn, who was still in Berlin and was presumed to be working on Germany's nuclear project. Meitner had reported to British agents that Hahn, while visiting her in Stockholm in October 1943, had mentioned that he saw "no practical utilization of fission for many years to come." No serious push for a nuclear weapon was being made in Germany, Meitner assured them. A similar conversation between physicist Paul Scherrer and German chemist Klaus Clusius had been reported, in which the German had indicated that all real efforts toward uranium isotope separation had largely been abandoned. Also taken as evidence of an inconsequential German nuclear program was the fact that the codebreakers at Bletchley Park had never come across any communications that made reference to either uranium or to fission.

But for Groves and Lansdale, all this information was hearsay at best. It was also impossible to confidently rule out the potential that the British sources were being fed intentionally misleading information. By this time, "it was so clear . . . that [a bomb] was practical" that they "could not believe that brilliant people like Hahn and Heisenberg could not reach the same conclusions." Groves did not want to just rely on this British intelligence; he wanted his own men on the ground in England. In early 1944, Captain Horace Calvert was permanently assigned to the London office of the Alsos Mission.

Calvert's first priority was to determine approximately how much nuclear material the German scientists actually had at their disposal. Before the war, the world's primary source of uranium was being mined in what was then called the Belgian Congo by the Belgian mineral company Union Minière du Haut-Katanga. It was known that the company had amassed a stockpile of over a thousand tons of uranium ore, which before Germany's invasion of Belgium had been stored in a warehouse in Olen, just northwest of Brussels on the Albert Canal. In spring 1940 when the Germans entered the Belgian capital, they had confiscated this stockpile of ore, and it was believed that they had moved the majority of it to Germany.

However, the Belgian stockpile was not the only source of uranium that the Germans had access to. A lesser but still usable grade of uranium ore was being excavated from the Joachimsthal mines in Czechoslovakia. Calvert set up aerial surveillance of these mines, as well as of Auergesellschaft, the German chemical firm believed to be responsible for extracting and purifying the uranium, in an attempt to estimate the amount of fissile material that Germany was extracting. The aerial images of the tailing piles around the mines were examined with a microscope to measure any increase in their size from one week to the next. While it was clear that the mines and chemical plant were active, the images from these reconnaissance missions showed that the processing facilities in Germany were nowhere near the scale of what would be required to feed the uranium needs of an atomic program even close to the size of the Manhattan Project.

Calvert and his team in London, aided by Lansdale's intelligence network in the United States, painstakingly pieced together all the data they could find on the current state of German science. German scientific journals were scoured for clues, and Manhattan Project scientists who had studied or worked in Europe before the war were questioned for any insights they might be able to provide on their former classmates and colleagues. From the information he gathered, Calvert compiled a dossier complete with profiles of people of interest, and a long list of possible military, industrial, and academic targets. All that remained for the Alsos Mission was to wait for access to mainland Europe, and by early 1944, plans for Operation Overlord (the code name for the Battle of Normandy) and the Allied invasion of the Continent were well underway.

Unbeknownst to the majority of the thousands of Allied commanders and infantry who were preparing for D-day, this bloody and grueling military offensive would also mark the first real nuclear gamble in the new Atomic Age. While all intelligence gathered by both the American and British agencies suggested that Germany most likely did not possess any explosive nuclear capabilities, there was no way for Groves and the top Allied commanders to be absolutely sure. Already well entrenched in a standoff with the Soviets to their east, the Allied invasion of France would force the Nazi army into a two-front war. Hitler had long been promising to reveal a new and terrifying V-2 weapon. Surely, if German scientists had developed a nuclear explosive, D-day would present them with an obvious opportunity to unleash it.

But General Groves was also keenly aware that an atomic bomb as the Manhattan Project scientists had envisioned it was not the only possible deadly use of radioactivity that the Nazis might have at their disposal. In addition to the iron obstacles and scattered land mines, it was possible that Allied troops landing on the French beaches might encounter sand that had been laced with highly radioactive materials. While the engineering and industrial work required to produce a nuclear weapon is immense, as the Manhattan Project's ceaseless work was demonstrating, production of plutonium was less complex, along with the immensely radioactive by products of its manufacture, and unprotected exposure to this material could prove devastating to the incoming soldiers. If the German scientists had succeeded in building a nuclear reactor by this time, and were using it to generate plutonium, they might well be able to use radioactivity to deadly effect, even without a nuclear explosion.

The Allied troops waiting in the wings on the British coast for the impending invasion of France knew nothing of the possible radioactive danger that might be waiting for them on the other side of the English Channel. General Groves understood that the invading soldiers risked encountering whatever form of nuclear weapon the German scientists might have concocted, but he also worried that sharing his concerns too broadly could compromise the security of the Manhattan Project. The decision was made to allow D-day to go ahead as planned without warning any of the commanders about the invisible dangers that their men might encounter. Protecting the secrecy of the American nuclear program, Groves reasoned, was worth the risk. "But," he later recalled, "it gave the handful who knew about it some bad hours" in the days leading up to the invasion.

Despite the need for secrecy, Groves felt that finding out what, if any, nuclear weapons the Nazis chose to use on the landing Allied soldiers was also important. So scattered among the landing boats that plowed their way across the English Channel on June 6, 1944, was a small group of men outfitted with radiation detection equipment. Code-named Operation Peppermint, these men had been trained in how to use and read the X-ray film strips and ionization chamber detectors that they carried, though they were not informed of the significance of their readings. Groves waited as Allied troops spilled out onto the beaches, and the Peppermint team radioed back the measurements on their meters: they weren't picking up anything at all.

The lack of any radiation-based defensive action by the Germans on D-day was not just a lucky break for the Allied forces—it was yet another hint at the larger reality of the German nuclear program. It implied that the German program was very possibly as far behind the United States in their nuclear capabilities as the limited intelligence suggested. Still, no one was willing to rule out the possibility of Germany producing a functioning nuclear weapon until there was positive proof to the contrary. Boris Pash and his Alsos team were waiting in their London base to enter Europe and find out for sure.

9 | THE HUNT FOR FRÉDÉRIC JOLIOT-CURIE

French Coast
August 1944

A FEW WEEKS AFTER D-day, on August 5, 1944, Lieutenant Colonel Boris Pash finally received Alsos's first objective on the newly liberated French coast: apprehend renowned French physicist Frédéric Joliot-Curie. Joliot-Curie, who had married into the famous Curie dynasty, and who with his wife, Irène, had won the Nobel Prize in Physics for discovering artificial radioactivity in 1934, was widely acknowledged to be one of the best scientific minds in Europe. In 1938 the Joliot-Curies had conducted experiments on uranium that replicated the results obtained by Hahn and Strassmann in their early fission experiments. On the heels of the publication of Meitner and Frisch's paper, Frédéric was able to again replicate Hahn's experiments and confirm Meitner's predictions about the release of energy and neutrons following a fission event, validating the possibility of a chain reaction.

Finding and capturing the French scientist became the first high-priority target for the Alsos Mission. Not only did the French scientist have extensive and intimate knowledge of nuclear physics, but also his close affiliation with France's Communist Party spurred (unfounded) concerns that the famous scientist harbored Soviet sympathies and may be working as an informant. The Joliot-Curie laboratory also housed one of Europe's only working cyclotrons—the same type of particle accelerator under construction on a colossal scale at the University of California, Berkeley, for America's own nuclear program. Intelligence reports from Paris suggested that for several years, Joliot-Curie's lab was being occupied and his particle accelerator used by physicists from Germany. Whether Joliot-Curie was a willing collaborator or if he had been shoved aside by the Nazis who invaded his laboratory remained unclear. But Groves and Pash were willing to bet that if there was anyone in France who knew what the Nazis' nuclear scientists had been working on, it would be Joliot-Curie.

Pash received a telegram at the Alsos headquarters in London that indicated that Joliot-Curie had been spotted a few weeks before at his summer home on the Brittany coast. Allied forces were quickly securing large portions of the French coastline, and Brittany would soon be accessible. Pash was to lead a team to this area and, if possible, take the physicist into custody.

Three days later, Pash, along with Gerry Beatson and Carl Fiebig, boarded a courier plane that took them across the channel and deposited them on Omaha Beach. The seafront still bore fresh marks from the intense fighting that had taken place there only weeks before. The small team acquired a jeep and moved south along the coast of France over rough mine-damaged highways, dodging the bombed-out tanks, both German and American, that dotted the road along the path being relentlessly carved through France by General Patton.

Following Patton's progress south, Pash, Fiebig, and Beatson slowly made their way to L'Arcouest, the quaint fishing and resort town on the coast where the Joliot-Curies had their summer cottage. They arrived at the coastal haven, machine gun fire from the front booming in the distance, to find a single street, lined with a few stores and a wine cellar. The throughfare was capped at one end by a small church and cemetery, and at the other by the rocky shoreline where an isolated German unit had built up fortifications and were refusing to surrender.

Pash asked some of the locals in the town if they knew of Joliot-Curie or where his home was. Borrowing a British army map, a man who claimed to be a cousin of the renowned physicist pointed out a small road below the village on the shore. Not only was the spot in the middle of the effective no-man's-land between the Allied troops and the entrenched Germans, but also the map indicated that this area was likely heavily mined.

Not to be deterred in pursuit of their first high-priority target, Pash and Beatson quietly made their way to the inlet of the road up to the Joliot-Curie house. When they arrived, they found what was left of a narrow path, over-run with bushes and other vegetation. To avoid booby traps, the pair crawled the distance to the house on their stomachs, feeling for trip wire and inching forward through the brush. Their movements did not go unnoticed, and their precarious situation was further complicated by a sudden barrage of gunfire, first from the Germans and then returned by the Americans, that was sent flying over their heads.

When the two men finally reached the entrance of the small stone house, they found the door ajar. As they slipped inside, guns still firing over the tangled yard behind them, they looked around at barren rooms, lit by the gray sea light streaming in through the windows onto the wooden floor. The house had been deserted. Not only had furniture been removed, presumably for firewood, but the drawers were all emptied, and the walls were left completely bare. Joliot-Curie was not here—the first Alsos Mission in France had proven to be a wild-goose chase.

With nothing to show for their troubles, the pair repeated their slow and treacherous journey back to the road, this time in reverse. They snaked their way back to the American line, arriving just in time for the surrender of the German unit that had been firing over their heads hours before.

Pash and his men returned to London to await the fall of Paris, when they would get another shot at locating the elusive French physicist.

10 | PARIS

France
August 1944

THROUGHOUT THE SUMMER OF 1944 the core membership of the Alsos Mission into Germany assembled at their London headquarters inside the US embassy. Alsos's scientific chief, Samuel Goudsmit, arrived in England on D-day, just as Allied troops were making their way onto the Continent. As he walked from the Mount Royal Hotel near Hyde Park to the Alsos offices, he was flabbergasted by the damage to the city from years of German bombing raids. Piles of rubble lay where buildings once stood, and though the poor East End neighborhoods of London had taken the brunt of the beating, there wasn't a street in the city that didn't show some mark of the war.

While the heaviest bombing campaigns from the Germans had ceased in 1941, London was again subjected to German attacks by air in 1944 as the Alsos Mission was planning and waiting. But this time, instead of heavy Luftwaffe bomber planes soaring over the city and unleashing destruction below, the Germans were making use of a new type of weapon, the V-1 rocket. Launched toward England from the French coast, about two thousand of these unmanned flying explosives, sometimes called buzz bombs, fell around London that summer. The incoming projectiles would trigger alarms at the US embassy, signaling to those inside to take shelter in the basement. These warnings were seldom heeded by the Alsos members, who were all eager to enter the fray. Instead, when the alarm sounded, the Americans would make their way up to the building's roof to hear the bombs, screeching as they fell, and to watch the fiery explosions that lit up the darkness of the blacked-out metropolis. Reginald Augustine, a lanky young lieutenant with short dark hair and a long neck who had been selected for the mission for his fluency in French, arrived in England midsummer. He recalled that during those weeks in London he "never even saw the basement of the Embassy."

In mid-August, word finally came down from Washington that Alsos would be heading into France. On August 19, this time accompanied by Horace Calvert, Pash traveled from London down to Portsmouth, and hitched a ride on a patrol boat across the channel. The bulk of the Alsos members would make the crossing in the coming weeks after Paris was liberated. Goudsmit, not yet fully prepared for the grim realities and discomfort of the front, told Pash that he would meet the team in France once adequate accommodations had been arranged.

Once in France, Pash and Calvert met up with fellow Alsos members Carl Fiebig and Gerry Beatson, who had remained in Europe after the failed Joliot-Curie mission and had been tasked with finding an American enlisted man, preferably one who could speak French and who had been issued a gun, to join them. Beatson had selected a clerk typist named Nathan "Nat" Leonard. A high school French teacher from upstate New York, Leonard had tired of his desk assignment and was eager for the chance to see some of the war.

Bringing Leonard into the mission also resulted in the incorporation of one additional team member. Leonard had come across a puppy in the French countryside a few weeks before and had been taking care of the small black-and-white dog. Pash, only feigning reluctance, allowed the puppy to join the crew on the condition that he be named Alsos, which would provide a convenient explanation for the mission's conspicuous name.

Beatson had arranged for two jeeps to carry the small outfit. Duffel bags stuffed with extra gear and K rations were strapped to the front grille and sides of the vehicles, a spare tire peeking out the back. The windshields of the jeeps were left in place, though they could easily be detached and strapped to the hood should the team encounter enemy fire.

Alsos's primary objective in France remained locating and apprehending Joliot-Curie, who was almost certainly to be found at his laboratory at the Collège de France in Paris. Pash's plan was to attempt to locate and join up with T-Force, an unattached conglomeration of intelligence-gathering units from several of the Allied countries, commanded by Colonel Frank Tompkins, and to move with this unit into the French capital.

The Alsos jeeps sped off through the French countryside, navigating narrow dirt roads shaded by roadside trees. Vast level fields nearly ready for the late-summer harvest surrounded them on either side. Tracking down the large but elusive T-Force was not an easy task, but Pash and his men eventually caught up with them in a small village just southeast of Paris. Finding Colonel Tompkins in his temporary office, Pash was distressed to learn that Charles de Gaulle, the head of the exiled French government, had requested that French troops be allowed to enter Paris before any of the American or British units. The Allies had agreed, and as a result T-Force would not be among the first units to enter the city. Recalling his experience operating behind the front in Italy, Pash felt that Alsos simply could not wait—they had to be among the first into the city if they had any hope of getting to Joliot-Curie and his laboratory before anyone else.

Pash requested a letter from Tompkins giving his unit special permission to join in with a French armored division that was set to lead the column of French troops into Paris. Hours before the sun rose the next morning, Pash and his men were speeding across the fields toward Highway 20, where in the early morning light they found a line of French tanks idling in the road. This was the division that they were supposed to accompany.

The Alsos jeeps idled for a while alongside the tanks until, irritated as always by the apparent inertia of the military bureaucracy, of which Pash believed the French to be particularly guilty, Pash made the brash decision to move ahead alone. In an instant, before anyone could stop them, the Alsos jeeps quickly cut out of the line and sprinted toward Paris on their own.

Their path took them through a series of villages on the outskirts of the city, where their small convoy was the first these towns had seen of the advancing Allied troops. As they passed through each hamlet, they were met with immense crowds lining the streets, singing "La Marseillaise" at the top of their lungs, and covering their jeeps with flowers until, according to Pash, they "resembled the prize floats in the Pasadena Rose Bowl Parade on New Year's Day."

At the Porte d'Orléans, the Alsos jeeps were brought to a stop just outside the city lines. There were still more than a few German units lurking throughout the city, and Alsos was in no way prepared for a fight in the streets. Pash also worried that if the team independently violated the high-level agreement

Allied armor enters Paris. Parisians mobbed the column, threw flowers and kisses, and even climbed aboard. *AIP Emilio Segrè Visual Archives, Gift of Michael Thurgood Haynes and Terry Thurgood*

made among the Allied forces about entering the city, even with the special high priority given to the mission, they might find themselves in a serious amount of trouble. When the French tanks finally rolled up behind them, perhaps spurred into motion by the reports of an American unit making haste toward the city, the Alsos jeeps quietly slipped into their ranks, officially entering the newly liberated Paris on August 25 at 8:55 AM. Pash's jeep was the sixth Allied vehicle, and the first American troops to reenter the city as the Germans pulled out.

Exuberant Parisians poured out of their homes to welcome their liberators. As the tanks and jeeps wound their way through the iconic tree-lined city streets, they were once again showered with flowers as French flags unfurled from the windows around them. Especially enthusiastic celebrators climbed up onto the tanks, riding them victoriously through the city.

But the retreating Nazis did not go quietly. As the advancing column approached the Hôtel des Invalides, a barrage of machine gun fire broke through the celebrations. The tanks and jeeps scattered, though the civilian revelers remained largely undeterred.

By four in the afternoon, the Alsos vehicles finally arrived outside the front gates of the Collège de France. They found Frédéric Joliot Curie waiting inside with a few of his students. No German scientists were anywhere to be seen.

The famed physicist and his assistants were all sporting armbands emblazoned with the French flag, markers of the French Resistance. It turned out that in the final weeks of the occupation, very little physics work was being undertaken in Joliot-Curie's lab. Instead, the group had been furiously constructing small homemade grenades and Molotov cocktails for the French Resistance fighters who had been using these homespun weapons to destroy German vehicles around the city. Word that Joliot-Curie's lab was the source of these devices had started to spread around Paris, and Joliot-Curie had begun to fear he might be a target of revenge from remaining Nazi loyalists. He was relieved that the Americans had found him first. His new captors, on the other hand, while glad to have finally caught up with the French scientist, were more than a little alarmed at the amount of explosive material they found piled up in cabinets and on benches around the laboratory halls.

That evening, Joliot-Curie's students and the Alsos men cooked a makeshift dinner together in the laboratory and celebrated the liberated city with two bottles of champagne, drunk from empty chemistry beakers, that someone had stashed in one of the Alsos jeeps during the mad dash across the French countryside.

Pash and Joliot-Curie stayed up late into the night, as gunfire popped and crackled through the city streets, discussing Joliot-Curie's interactions with the German scientists. He didn't know much, he admitted from the start. He had been contacted early on by two high-ranking German scientific officials—a Professor Schumann and someone that the Alsos men had not heard of before: Kurt Diebner. The Nazis had initially wanted to move the cyclotron in the Joliot-Curie lab, along with the rest of his equipment, to Germany, but had later settled on sending their scientists to Paris instead. The takeover of his lab made Joliot-Curie appear, at first blush, to be a collaborationist, when in fact he had spent the last several years doing everything he could to avoid, or when possible, stall his new lab-mates. In fact, days before the Nazi occupation of the city, Joliot-Curie had dispatched his colleagues, Hans von Halban and Lew Kowarski, to England. The pair narrowly escaped Nazi commandos in Norway, arriving in the United Kingdom along with all of their early data on building nuclear piles as well as France's entire two-hundred-liter supply of heavy water (the largest stock in the world at the time) to aid in Britain's Tube Alloys program. But unfortunately for Pash and Alsos, in the way of information on Germany's nuclear research, Joliot-Curie had very little to offer.

The next morning, stepping outside the Collège de France, the Alsos team found a city slowly coming back to life: "There was feverish activity everywhere to clear away barricades and carcasses of dead war-machines." Unlike London, or even the small villages surrounding the city, the Nazis had largely spared the French capital. Most of the architecture remained unharmed, and as the cafés started to reopen, Paris quickly began to resemble the City of Lights from before the war.

Colonel Tompkins and his T-Force were located at the Petit Palais between the Champs-Elysées and Pont Alexandre III. Borrowing their communication equipment, Pash sent word back to London and Washington of their successful entrance into the city, and of the capture of Joliot-Curie, who was to be escorted to London, where he was held and questioned for a time. The successful capture of their first major target was welcome news to those keeping tabs on Alsos's progress back in Washington.

Over the next few days, the Alsos team began setting up their own headquarters nearby, on Place de l'Opéra, while they waited for the remainder of the Alsos members, including their scientific team, to arrive.

Goudsmit's arrival on the Continent on August 27 was not as smooth and comfortable as he had hoped. The shuttle plane from Britain had taken off later than planned because of fog, and when Goudsmit had arrived in Cherbourg, France, he found that no arrangements had been made to collect him. In fact, he had missed the communications from Pash and had no idea where in France the rest of the Alsos team might be. Thinking through his predicament, he commandeered a military vehicle and driver that had been intended for a senior officer who had missed the flight from England. Goudsmit spent the next two days directing the poor driver from base to base around western France, trying to locate some trace of his team.

As he drove, he saw evidence of the war at every turn. Tanks and destroyed vehicles littered the sides of the road, and huge swaths of land were marked with boards and flags indicating mines that had not yet been cleared. He had seen some of the destruction of the villages along the French coast in aerial reconnaissance images, and from the plane window as he flew in, but seeing the carnage at ground level was another thing entirely. Whole villages had been reduced to rubble in some areas, and the citizens were now undertaking the gruesome process of digging out the dead.

The persnickety scientist found himself spending his first few nights of active duty in abandoned schools that had been converted to refugee shelters, with little access to water, and horrible bathroom facilities. Finally, at the end of the second day of wandering, he received a tip that the Alsos Mission was convening at their new headquarters in Paris. He set off for the city the next morning.

Meanwhile, on August 29, Pash, unaware of Goudsmit's difficulties, was shown a startling message from the American forward base at Valognes, requesting that the American commander in Paris "arrest and hold one Samuel Goudsmit of The Alsos Mission, for absconding with a government vehicle." It seemed that Goudsmit had never officially received permission to use the car that he had been traveling in, much less take it to Paris. Pash and his men sprang into action, desperately trying to find Goudsmit before the military police could get to him. But no one had any idea where Goudsmit might be. None of the checkpoints between the coast and Paris reported any sign of him.

A few hours later, a lone army jeep rolled into Paris carrying the scientist, who was not at all concerned that he was the subject of an army-wide manhunt. Instead, he was eager to get cleaned up and down to business. Exasperated with this unflappable new side of his colleague, Pash asked Goudsmit if he thought he would be able to adequately direct the mission from jail. Fortunately, Pash was becoming quite adept at using Alsos's high profile and priority status as a means of smoothing over tense situations, and soon Goudsmit was cleared and the Alsos Mission was able to continue their investigations.

11 | BELGIUM

September 1944

IN EARLY SEPTEMBER ALLIED FORCES began making their way north toward Brussels. Belgium was of particular interest to Groves, and therefore the Alsos Mission, because Belgium's colonial presence in Africa had given them access to what was at that time the largest uranium mining operation in the world. The Belgian company, Union Minière du Haut-Katanga, had stockpiled thousands of tons of uranium ore. When the German forces occupied Belgium in 1940, they gained access to this massive amount of material. But there remained a possibility that some of the uranium stock might still be found in Union Minière's warehouses. Locating and capturing any remaining uranium ore became the Alsos Mission's next top priority.

Before Alsos could head north, Pash went to see Colonel Bryan Conrad, assistant chief of staff for intelligence (G-2) in Europe, in order to obtain clearance for his team to move into the Belgian capital. The area around Brussels was entirely under British military control, and while Alsos was well acquainted with British intelligence, they had not yet had any real contact with the British Army. Rather than rely on paperwork to clear their way, Conrad, bored with the endless office work behind a desk and eager for an excuse to get out into the field, offered to personally accompany the mission members and ensure a smooth operation. Walking out of his Paris office with only a pistol and couple of clips, and telling his secretary he would back later, he hopped into Pash's jeep, and they headed toward Brussels.

The Alsos team left Paris through the Porte de Clichy and traveled north toward Amiens. They drove the 190 miles to Brussels along bumpy, crater-filled roads, lined with bombed-out factories and torn-up railways. The small convoy was almost blocked when they approached the Oise River to find that the only bridge had been destroyed. A solution was soon discovered when they spotted a local farmer who was using a makeshift raft as a ferry: "With the help of his raft and our cigarettes, we moved the jeeps across the Oise."

When the Alsos team entered Brussels later that day, they made their way through the jubilation and chaos of the newly liberated city to the offices of the Union Minière headquarters. Told that they would not be able to enter the offices or speak with any of the Union Minière officials until the next morning, the team spent the evening exploring the city and celebrating with the ebullient inhabitants. A walk through the central zoo revealed a particularly unusual sight. The zoo's enclosures, devoid of their animal inhabitants, all having been killed and some even eaten during the occupation, were now being used as the city's makeshift prison for Nazi collaborators.

The next morning at the Union Minière offices, the Alsos team was able access records that gave the Americans their first real accounting of the Belgian stores of uranium. The Nazis had slowly been removing ore from Belgium since the occupation in 1940, and in total they had moved about 1,100 tons of the material across the border into Germany. However, it appeared that about seventy tons of uranium ore were still located in the Olen warehouses. This region, to the north of the capital city, was still an active combat zone—Alsos would have to wait for the fighting to die down before trying to locate the remaining ore.

From the Union Minière documents the Alsos team also learned that in early 1940, in an attempt to keep the material out of Nazi hands, another eighty tons of uranium ore had been hastily shipped into southern France. The whereabouts of the seven railcars that carried the barrels of powder to safety were unknown.

The German chemical company Auergesellschaft had a refinery in Oranienburg, in northeastern Germany, that was thought to be the likely destination of the uranium shipments from Belgium. The records found at Union Minière also linked Auergesellschaft's operations to Terres-Rares, a formerly Jewish-owned rare mineral distributor out of Paris that had been taken over by Germans during the occupation. Pash made a call back to Alsos's headquarters in Paris and relayed this information to Sam Goudsmit, who rushed over to Terres-Rares's abandoned offices to investigate.

Goudsmit's efforts revealed records from the French company indicating that in addition to the uranium shipments from Belgium, an enormous amount of thorium had been moved from France into Germany over the previous few months. While not naturally fissile itself, thorium can be transmuted to fissile ^{233}U through irradiation in a nuclear reactor. Goudsmit feared that these shipments might indicate that the Germans were trying to build a thorium-powered explosive.

From the papers found in Paris, Goudsmit identified a Mr. Petersen, the German who had been placed in charge of Terres-Rares after the occupation and, they believed, had already disappeared behind the German border, as the primary person of interest. Further digging revealed that Petersen's secretary, a Mademoiselle Wessel, was likely at her parents' house in the town of Eupen on the Belgian-German border. Anxious for more information, and still waiting for access to the Olen warehouses, Pash characteristically took the matter into his own hands, and the Alsos jeeps set out again, this time heading east.

When they arrived in Eupen they found a small two-story family home. Pash knocked on the door and a young woman answered. Seeing Pash's American uniform, her face suddenly went white with panic, and she immediately slammed the door in his face. Pash knocked again, more forcefully this time, and the door was answered, this time by an older gentleman who was apparently the young woman's father. Pash, Fiebig, and Beatson entered the Wessel home to find the whole family nervously huddled in the dining room.

With Fiebig translating to German, Pash told the family to remain where they were while they searched the house. Fiebig and Beatson moved through the home and up the stairs, checking each room until they came across a conspicuously locked closet. Pash called to the father to come upstairs, and after hedging a moment, the old man admitted that a friend of his daughter's, a man, was behind that locked door. He insisted that they had only lied to spare her the embarrassment of the suggestion of an improper relationship. After hearing this conversation through the door, the enclosed individual decided that the jig was up, and the door was unlocked from within. To Pash's amusement, the man who emerged identified himself as Mr. Petersen from the Paris company. Petersen, along with Mademoiselle Wessel, who was apparently more than just Petersen's secretary, became Alsos's first real prisoners. Pash called Sam

Goudsmit to tell him that they had gotten their man. Goudsmit was briefly confused; hadn't they gone to Eupen looking for a woman?

Pash brought the pair back to Paris. In an empty hotel room, the subjects were seated facing the windows, as the Alsos team prepared for their first interrogation with great fanfare. Samuel Goudsmit later recalled, "We put on our Sunday uniforms and Colonel Pash put on as many ribbons and medals as he could find." But all the pageantry and pressure amounted to nothing—Petersen was unaware of the significance of thorium for atomic weapons. The rare metal, it later turned out, had been transferred to Germany not for scientific purposes but as part of an enterprising postwar business plan to manufacture and market thoriated toothpaste.*

The Belgian uranium, however, had no such unusual alternative purpose. As soon as the fighting had died down enough for them to gain access, the Alsos men traveled to Olen in two armored cars, with Pash's uncovered jeep leading the way. The uranium was purported to be located in a warehouse complex of boxlike buildings several stories high that occupied the buffer zone between British forces on one side of the Albert Canal, and the Germans, just a few hundred yards away on the opposite bank. As the Alsos team got out of their vehicles, the Germans made it known that they were watching from their posts, taking a few potshots at Pash and his men across the water.

Dodging fire along the bank with the whining of the mortar projectiles being flung over their heads at the British line behind them, Pash, Fiebig, Beatson, and Augustine split up and fanned out through the buildings. They wanted to appear to be checking all of the warehouses, not just the one containing uranium, on the off chance that anyone watching them might be trying to infer what it was they were hoping to find.

The warehouse manager directed them to one of the buildings, where they found hundreds of small but heavy barrels stacked up against the back wall. As this was much more material than the four men would be able to move on their own, they decided to leave it in place with strict instructions to the manager to not let anyone else near it. Once back in Brussels, Pash placed a series of phone calls, and soon arrangements were in place to evacuate the barrels to London.

---

* The company even had a proposed slogan: "Toothpaste with thorium, for *radiant* white teeth!"

12 | UNOCCUPIED FRANCE

Southern France and Paris
October–November 1944

RETURNING TO PARIS FROM the Belgian expedition, Pash reported to the new Alsos headquarters, now located on the sixth floor of the US Navy office building at 9 Rue de Presbourg. Pash was pleased with their new space, which they would occupy throughout the remainder of the war. But the new offices were not the only upgrade that the mission had made while Pash was away. A short distance down the road from headquarters was the Royal Monceau hotel. Upon entering the swanky establishment under a glowing red awning, Pash was given a key at the front desk to one of the hotel's most luxurious suites that included a well-furnished sitting room looking out over Avenue Hoche, and a gleaming bathroom complete with a French-style tub standing imposingly on its claws. Much to Pash's delight, and no doubt Goudsmit's relief, this sumptuous establishment would be the central home base for all the mission members while in Paris. The restaurant downstairs became the site of many late-night, cognac-fueled planning sessions in the coming months as Alsos slowly worked their way to their final objectives.

But Pash was unable to enjoy his new accommodations for long, as the team set to work tracking down the missing eighty tons of ore that the Belgians had shipped into southern France ahead of the German invasion. Pash recalled, "The Pentagon kept firing advices at me to the effect that our atomic project sorely needed the uranium supply and that I was expected to come up with it immediately if not sooner."

The few records that they had been able to find in Brussels hinted that the train cars carrying this missing uranium might be located in or near Toulouse, only about a hundred miles from the French border with Spain. This presented a problem. This southern region of France, south of the Loire and west of the Rhine, was a no-man's-land, occupied only by the French Forces of the Interior (FFI): partisan fighters mostly operating in small, loosely connected groups, each defending its own hometown. No Allied troops were allowed to enter this region without special permission, but, of course, Pash was able to leverage Alsos's special status and obtained the necessary paperwork. However, the lack of Allied presence in the area meant that his team would be totally isolated and at the mercy of the partisan fighters, who had a reputation for shooting first and asking questions later.

In the first days of October, six Alsos men set out in their customary two jeeps, a large American flag attached to the side of the lead vehicle. With them was Major John Vance, a chemist from the Manhattan Project, who had crossed the Atlantic with Geiger counter in hand, to help ensure the successful removal of any uranium they might find.

The Alsos convoy passing through a road barricade in southern France.
NARA II, RG 165, Box 161, Alsos Mission Album

Navigating southern France was difficult and slow. In an effort to keep out the German invaders, the FFI had blacked out road signs all over the region. To control the influx of any military forces trying to enter a town, the partisans had constructed roadblocks made of logs stacked four feet high, and set at a diagonal across the roadway, leaving only a small gap of open road, just wide enough for an incoming vehicle to pass through. Several of these blockades were arranged on alternating sides of the road like a giant zipper, forcing an incoming vehicle to slow to a crawl as it zigzagged its way past the obstacles, and allowing plenty of time for the armed guards to determine if the intruders were friend or foe.

In most towns they passed through on their way south, the Alsos team were the first Allied troops that any of the residents had seen since the French and British armies had pulled out of mainland Europe in 1940. To ease their movements around this region, and for help in locating the uranium, a letter of introduction had been provided to Pash in Belgium by officials at Union Minière that was to be shown to a Frenchman named André Polette, a high-ranking official in a French firm affiliated with the Belgian mining company. Alsos located Polette at his family home, Château la Meynardie, near the tiny haven of Coquille. The stately house was occupied by Polette, his wife, and their three daughters, all of whom were thrilled by their new visitors, and by the supplies of sugar and coffee they brought with them. Polette was happy to assist Pash and his men, and even offered the use of his home as a base of operations.

With Polette's help, Pash acquired paperwork that would allow Alsos to inspect and remove any materials that they identified as belonging to Union Minière from the train depot and armory in Toulouse. A few days later, the team made their way into the city at dawn, driving through streets lit with a warm pink glow from the rising sun reflecting off the terra-cotta brick buildings. Their early-morning appearance at the gates of the train depot startled the manager, who, still half-asleep, opened the gates and allowed them to enter. Pash kept the manager busy in his office, peppering him with questions, while Vance and Augustine snuck away to search the depot warehouse unencumbered.

The old wooden structure was filled with haphazard assemblies of crates and containers, and the pair navigated their way around the piles until they found a sizable collection of small barrels, each marked with Belgian shipping labels. As they approached the barrels, the needle on Vance's Geiger counter dial began to jump.

After the team located the uranium, the next question became how to get these barrels of crushed rock out of Toulouse. Vance made a quick back-of-the-envelope estimation and guessed that the pile of barrels accounted for around thirty of the eighty tons of uranium that had been sent into France at the start of the war. This was far too much material to carry out in their jeeps, so Pash hastily returned to Paris to work out the details of an alternative plan.

Pash was granted the temporary loan of an elite truck company from the Red Ball Express, the Allied supply line that had been set up between the coastal ports and the fighting fronts in France and Belgium. While he was negotiating the loan of the trucks, Beatson and Augustine, who were still in southern France, called their commander to let him know that Alsos's interest in the region had not gone unnoticed. Their sudden appearance at the train depot had spawned a fast-spreading rumor that the barrels must be filled with gold. So in addition to the trucks, Pash also requested a combat squadron to accompany them in case the French citizens had any last-minute objection to the mysterious barrels being removed by the Americans.

On October 10, an exuberant Pash, leading a calvary of massive armored trucks, rolled his way back into Toulouse. The show of force worked as intended, and the impressive line was allowed to enter the depot unopposed. The expert drivers backed the enormous vehicles into place in perfect formation in front of the warehouse door. Loading the material did not take long, though the men, who had no idea what it was they were there to retrieve, were shocked to find that the small barrels they were carrying out were almost unfathomably heavy. The thin film of yellow dust that the barrels left on the ground and on their gloves only served to fuel notions of their supposedly auriferous contents.

Leaving Toulouse, the convoy headed straight for the port of Marseille on the southern coast of France, where the contents were to be transferred onto a US Navy transport ship. During loading, one of the barrels slipped from the crane and fell into the sea, and navy divers had to be called in to recover it before the ship set sail for Boston. The crossing took two weeks, and when the ship docked in Boston Harbor, Augustine, who had been sent along with the barrels to ensure their safe arrival, found Major Vance, who had rushed back by plane, there to greet him.

Between the stores of ore that had been recovered in Olen and the additional thirty tons found in southern France, the lion's share of the uranium that was thought to be in Europe, outside Germany itself, was now safely on its way to the United States. While Alsos continued unsuccessfully to search for the remaining fifty tons of uranium ore that had been shipped out of Belgium in the early days of the war, as they waited for Allied forces to progress east, the team had little to occupy their time.

For Pash, Goudsmit, and their men, the autumn of 1944 was largely spent lounging in Paris cafés and theaters, the calm occasionally punctuated by a wild-goose-chase investigation. While Alsos was still unable to enter the regions where evidence of nuclear research was most likely to be found, anxious scientists back in Washington hoped it might be possible to identify the presence of nuclear research activity through other means. After Allied forces accessed the Netherlands, Goudsmit received a missive at Alsos headquarters from the Pentagon requesting that samples of river water from the Rhine be collected and sent back to the States for analysis. The Rhine flows northward through much of central Europe. Any large-scale nuclear reactor requires a large body of water for cooling. If such a facility existed in Germany and made use of the Rhine for this purpose, the chemists in DC hoped they might be able to detect some sort of nuclear signature in the water that might give some clues about the relative scale of Germany's nuclear program.

Two of Goudsmit's scientists were dispatched to the Rhine to collect the requested samples, which were bottled up for shipment to the United States. Almost as an afterthought, as a surprise gift for their stateside counterparts, a particularly nice bottle of French wine, which had become a scarce commodity in America during the war, was slipped into the crate carrying the samples with a note attached that read, "Johnny, give special attention to this."

Goudsmit and his men didn't give the matter another thought until they received a memo from Washington a few weeks later that read: "Water negative. Wine shows activity. Send more. Action." Goudsmit was initially tickled that their little gift had apparently been so well appreciated. His amusement quickly turned to annoyance, however, when it became clear that the request was no joke. A rapid series of further radiograms made it clear that the wine had not been consumed as intended, but instead had been tested along with the water, and radioactive isotopes had been detected. Rather than assume (correctly) that these readings were the result of the mineral contents in the water used to grow the grapes, the overzealous Washington officials thought instead that these readings might indicate the presence of a secret German mountain nuclear laboratory.

Irritated that they now had to waste time on this futile mission, but with Washington giving him little choice, Goudsmit sent Captain Wallace Ryan and physicist Russel Fisher into the heart of the French wine region to collect more samples. The lucky pair spent a marvelous few days traveling from winery to winery collecting samples of the wine, grapes, and soil, and doing their own sampling of a fair amount of the suspect liquid along the way. When they returned, a huge box of wine bottles and jars were carefully labeled and prepared before being sent back to the United States for analysis. Goudsmit, who had a deep appreciation of wine, bemoaned that he hoped that it had not all been squandered on unnecessary chemical analyses. Whatever happened in Washington when they received the package, that was the last Alsos heard on the matter.

For the rest of October and well into November 1944, the Alsos team carried on in this manner from their base in Paris. They pieced together what little information they had found, most of which indicated that the German program was at least not as large as the American effort. "But," Goudsmit would later write, "we never knew anything reliable about it until we knew practically everything at once and that was after the fall of Strasbourg."

13 | STRASBOURG

France
November–December 1944

THE CITY OF STRASBOURG, often referred to as the "jewel of the Rhine," sits along the river that, after World War I, served as the boundary between France and Germany. During World War II, when the German invasion moved through the Alsace region of northeastern France, Strasbourg was one of the first cities taken, languishing under Nazi control until the final weeks of 1944, when French Allied troops finally reclaimed it.

The scientists of the Alsos Mission were interested in Strasbourg for several reasons. More than anywhere else in France, the Germans had endeavored to incorporate the city's university and industrial centers into the main fabric of the Nazi war machine. Strasbourg was the most likely place, outside Germany itself, for the Alsos team to get a glimpse of Germany's war-science programs. The Bugatti-Trippel automotive plant was thought to house a model of the German torpedoes that had been wreaking havoc on Allied ships trying to cross the Atlantic Ocean. And another car manufacturer in the area, Mathis, was producing Luftwaffe fighter planes, and purportedly contained research on what was arguably Germany's greatest wartime scientific development: the jet engine.

As for nuclear research, Strasbourg presented the first real chance for the Alsos Mission to investigate a German physics lab. Intelligence suggested that the old and prestigious University of Strasbourg had been built up to serve as a model of an idealized Nazi research institution. This included the transfer of prominent scientists in physics and other fields along with their work from within Germany to the university. In particular, reports indicated that two significant German nuclear scientists were likely to be found at Strasbourg: Rudolf Fleischmann and Werner Heisenberg's young protégé, Carl von Weizsäcker.

Pash received notification of the imminent fall of Strasbourg on November 22 and with a small team that included Leonard, Fiebig, and Beatson, the outfit made their way in two jeeps with supply trailers from their Paris base toward the border city. The 250-mile trip through recent battle sites was slow, and the jeeps had only made it as far as the town of Saarebourg by late that evening. Just north of this small town, Germans were launching a last-ditch defensive, and the city of Phalsbourg, which lay between Alsos and their target, was reported to be only tenuously held by Allied forces. The men decided to spend the night in Saarebourg rather than risk completing the journey through enemy fire in the dark. As an independent unit, Alsos was particularly vulnerable to being left behind if the Allied army was forced to retreat.

Beatson managed to locate a place for the unit to stay overnight in the basement of an abandoned two-story house. Huddled together in the damp, with two candles providing all the light they were willing to risk, and enemy artillery fire sounding in the distance, they consumed the K rations and coffee that stood in for a Thanksgiving dinner. For Pash, this second missed turkey marked a full year with the Alsos Mission.

The men took turns guarding the jeeps throughout the night and tried to get some rest. Around 3:00 AM Beatson came flying into the basement to wake the others up. A stray shell had fallen in front of the house, destroying one of their supply trailers. The supplies had thankfully been equally distributed between the two trailers, so the team only lost half of their rations and equipment. But rattled and wide awake from the close call, they decided to pack up and move out.

With fighting continuing to their north, Pash decided to try approaching Strasbourg from the south through the Dabo Pass, taking the jeeps through the crimson autumn hills of Alsace. Travel through the winding roads was slow and tense, and on more than one occasion the jeeps rounded a corner to find a German tank sitting in their path. Thankfully all were inoperable, and their occupants deceased—telltale signs of the recent fighting.

Reaching the Saverne highway, Alsos caught up to American and British troops making their way to aid the French forces in securing Strasbourg. The situation in the city was still nebulous, Pash was told, but he remained eager as always to find their targets as soon as possible.

Entering the city early that morning, the Alsos jeeps kept to the periphery of town, avoiding the narrow, cobbled roads in the center of the city. Their first stop was a heavily damaged apartment building where their first target, Rudolf Fleischmann, supposedly occupied two small rooms. When they arrived, Fleischmann was nowhere to be found, and the elderly woman next door insisted that she did not know where to find the scientist. After some pressing, she did finally provide Pash and his team with the address of one of Fleischmann's colleagues. Alsos loaded their jeeps with all the documents and communications that they could find in the small apartment before setting out to find this friend.

The man in question was found at his home and claimed to be entirely unaware of his colleague's whereabouts. The conversation seemed like a dead end until out of the blue the man asked if he would be allowed to go to the hospital the next day.

"Don't you feel well?" Pash asked, trying his best not to react to the unusual comment.

"It's just that I do a little of my work there," the man replied.

Taking the unintended hint, Pash had the man and his wife placed under house arrest and cut their telephone line to prevent them from making any alerting calls. That evening at the T-Force base, Pash and his men made plans to access the hospital first thing the next morning.

Pash's hunch turned out to be correct. A little intimidation and a few threats to the man sitting at the front desk quickly revealed that Fleischmann had a lab located in a separate building at the edge of the hospital grounds. The Alsos cohort rushed to the location that the receptionist had indicated on a map. Before entering, Pash posted two of his men at the door, instructing that people be allowed to enter, but not to let anyone leave. Inside the building they found a few dozen scientists and assistants wearing white coats, just like the hospital's doctors and nurses, going about their work, apparently unbothered by the fighting of the previous few days. As hospitals had largely been spared any involvement in the conflict, these men and women believed that they would be left alone.

Pash and his men had herded all of the laboratory workers into one room for questioning when suddenly a particularly tall man who had been indignantly slouching in a chair sprang up and began shouting at Pash and his men in German. Pash summoned up his own limited German to bark at the man, who he correctly guessed was Fleischmann, to sit down and not open his mouth again until he was spoken to, or he would regret it.

The Alsos Mission had captured their first Nazi nuclear scientist, but finding their second mark in Strasbourg proved a disappointment. The next day, when the Alsos team arrived at Weizsäcker's apartment, they found that Weizsäcker had had more forethought about the impending Allied invasion than his colleague. He had left town several days before and the Americans found his apartment entirely emptied, except for the large stove in the living room, which they found full of ash and a few unburnt corners of paper. Pash guessed that more scraps of envelopes and letters might have made their way up the chimney and had the whole unit taken apart. The charred remains of Weizsäcker's communications were placed in an envelope for Goudsmit to examine when he arrived in Strasbourg later that week.

Interrogations of Fleischmann and investigations of his labs initially appeared to be less illuminating than had been hoped. The physics facilities in Fleischmann's labs were not as advanced as the Alsos team had been led to expect—it was clear that whatever nuclear physics research might be happening in Germany, Strasbourg was not really playing any central part. Fleischmann himself continued to be combative and uncooperative, and remained in an Allied holding cell, much to Sam Goudsmit's discomfort.

"But he's a scientist like me. He's not a soldier or a—a criminal!" Goudsmit exclaimed to Pash. Pash assured his colleague that while Fleischmann was a scientist, he was also, in his loyalties and in his belligerent behavior, very much the enemy.

The two nuclear scientists were not the only Alsos targets in the newly liberated city. A few weeks earlier, as the Alsos team had been in Paris planning for their move into Strasbourg, Major Burnes (British), Captains Cromartic and Henze (US Army), and Lieutenant Hoffer (US Navy), members of Alsos who were focused on obtaining information about German research on bacteriological weapons, had identified a man named Eugen Haagen, who specialized in weaponizing viruses and was known to be residing and working in Strasbourg.

In Haagen's lab at the university they found a grim scene. The once-renowned medical scientist, who had helped to develop the world's first yellow fever vaccine, had been conducting horrifying and macabre experiments on prisoners sent to him from Germany's concentration camps. What the American scientists found most distressing about this discovery was that Haagen, until very recently, had been counted among their own ranks. Just years prior he had spent time working in New York City under a prestigious fellowship at the Rockefeller Foundation. In hindsight, his dedicated involvement with the German American Bund during that time might have served as a warning sign of what was to come.

The Alsos team set up their forward base in Strasbourg at the home of the now-incarcerated Haagen, which was well furnished and large enough to house most of the team. Pash set up his office in the expansive library, from which Augustine took it upon himself to personally liberate complete sets of the works of Goethe, Schiller, and Heine for his own bookcases back home.

When Goudsmit and fellow Alsos scientist Fred Wardenburg arrived in Strasbourg a few days later, they set to work poring over the documents and charred scraps collected from the two German physicists' apartments. They spent hours hunched over papers, muttering quietly to themselves late into the evening, until suddenly Pash heard Goudsmit exclaim, "We've got it!"

"I know we have it. But do they?" Pash lazily replied.

Sam Goudsmit beamed, "No no! That's it. They don't!"

The documents that had made Goudsmit gasp with excitement were not official reports or detailed experimental plans, but casual letters sent between Fleischmann and his friends. While to most observers these communications might have appeared completely innocuous, they were filled with little offhand references to people and locations. And Sam Goudsmit knew what he was looking for. "To an outsider, a professor is a professor," Goudsmit later wrote, "but we knew that no one but Professor Heisenberg could be the brains of a German uranium project." These documents confirmed Goudsmit's initial suspicions: looking for the German nuclear program really meant looking for Heisenberg.

While the letters contained little direct information about the nuclear program itself, Goudsmit found numerous mentions of the famed German scientist. Goudsmit also found references to Belgian uranium and to the Auergesellschaft, which they had been surveilling for months from London. Following the breadcrumbs through the letters, he learned everything: about the limited scale of Heisenberg's reactor experiments in Berlin, and about a competing experiment at the military base of Gottow being conducted by someone named Kurt Diebner, one of the Germans who had visited Joliot-Curie's laboratory in Paris.

From what Goudsmit was able to discern from these letters, he was now absolutely sure that not only did Germany not have a nuclear weapon but they had no way of mass-manufacturing radioactive poisons, as had been feared before D-day. In fact, they had not yet even created a self-sustaining chain reaction. The sum total of the German program had not progressed any further than where the United States' own efforts had been in early 1940. Germany had no significant nuclear capabilities—Alsos had found the answer that they had been looking for.

What's more, they also now knew exactly where in Germany they would find Heisenberg and his small-scale experiments. The addresses and even the telephone numbers of the facility in the small town where Heisenberg was apparently conducting his experiments were printed clearly on the letterhead of some of the documents.

Pash reported their monumental discovery back to Washington as the "biggest intelligence bombshell of the war." He wondered what this would mean for the Alsos Mission. In large part, they had just accomplished their objective—they were sure without a doubt that Germany was not, nor was it likely to ever be, a nuclear threat. But if Pash had any ideas about slowing down the pace of the mission, or even of returning home for a while, they were quickly dispelled by the reply he received from Groves in Washington.

Groves did not doubt the veracity of Goudsmit's conclusions. In a memo to General Bissel, Groves wrote of the discovery, "This is the most complete dependable and factual information we have obtained. . . . Fortunately, it tends to confirm our conclusion that the Germans are now behind us." Nonetheless, letters and gossip would not serve for absolute proof. And the Nazis' nuclear program, however embryonic, needed to be confiscated along with its scientists before the mission could call its task complete. Not articulated outright in these memos—but still very much implied—was the necessity that the American Alsos Mission capture the whole of the Nazi nuclear program, lest the Soviets get hold of Germany's nuclear scientists and materials first. Pash and his men were heading into Germany.

14 | HEIDELBERG

Germany
February–March 1945

GOUDSMIT'S DETECTIVE WORK IN STRASBOURG had revealed exactly where Alsos would find the majority of Germany's nuclear research. What had begun as a list of hundreds of potential targets had now been reduced to just a handful of locations. But getting to these locations would have to wait. In the final days of 1944 and early 1945, German forces were putting up a formidable fight in the Ardennes forest, along the borders between France, Belgium, and Germany. The wall of fighting effectively blocked Alsos from accessing any German territory.

Following their discoveries in Strasbourg, Pash was called back to Washington for a few days to provide a more complete report on their findings and on the progress of the mission more generally. The Pentagon had received several complaints from both American and British forces about the mission and its rapid and unceremonious movements around the front. While in Washington, Pash was given an earful about the headaches his methods were causing his higher-ups. But it was hard to deny that so far the unorthodox mission had been quite successful, and despite the reprimand, Pash noted, no one actually told him to change anything about the mission's tactics or pace.

There was also some concern and curiosity in Washington about how the mission scientists were coping. Pash, who had grown extremely fond of his academic teammates, assured the men at the Pentagon that the scientists were proving quite capable in the field, and that "they can work, scheme, bitch, finagle and ride a jumping jeep right along with the best of our field soldiers."

After delivering his reports, Pash returned to his unit in Paris, arriving just in time for a Christmas Eve party at the Royal Monceau. He had returned from the States prepared. His baggage, which had weighed in at one pound over the limit, was stuffed entirely with "Christmas cheer" for his men. To make room for these yuletide provisions, he had all his clothing sent back to Paris by mail.

The early weeks of 1945 were uncertain for the Allies as the German offensive in Ardennes continued. The Alsos base at Strasbourg had to be evacuated briefly when Allied control of the city became tenuous. But by February the tides had once again turned in the Allies' favor and the Allied advance into Germany, along with Alsos's access to German nuclear research, seemed imminent. With Strasbourg again secured, the Alsos base was reestablished as Alsos Forward South. Another command post, Alsos Forward North, was opened farther north in Aachen.

Despite their two new forward positions, Pash and Goudsmit decided to keep Alsos's primary administrative center in Paris because of the proximity of their headquarters to both European Theater of Operations, US Army, also in Paris, and to Supreme Headquarters Allied Expeditionary Force at Versailles. Pash later admitted that their luxurious accommodations in the city might have had something to do with this decision as well.

As new cities and possible targets began to open, the Alsos Mission's scientists would venture out from their Parisian post and make their way to one of Alsos's new offices in the field. From Alsos Forward North in Aachen, they were able to gain access to the German city of Cologne and investigate a handful of industrial and academic laboratories there. As expected, not much additional information was found beyond what they had already learned from their time in Strasbourg, though they did find the occasional interesting or useful scientific apparatus to confiscate.

But it was not only scientific intelligence that found its way into the Alsos jeeps. As they were packing up and preparing to leave one university lab in northern Germany, Pash was startled to see Jim Lane, one of the mission scientists, walk out the front door carrying an armload of objects that had nothing to do with science. His souvenir haul included a swastika flag, a ceremonial knife, and a camera, among a jumble of other items. As he loaded his loot into the jeep, he laughed at Pash's dumbstruck face: "You know I am not the only scrounger in this outfit!"

Pash had to smile at this comment because he knew exactly how true it was. As the first Allied unit on the scene at most of their targets, the Alsos men had their pick of the spoils of war to choose from. The scientists, in particular, were keen on collecting mementos. In one instance a few weeks later, Goudsmit returned from Celle in the northernmost reaches of Germany, where he had been investigating the Germans' barely nascent isotope separation program, with reams of white silk fabric streaming from the back of his jeep. The lab, it turned out, had been located on the grounds of a parachute factory. When he returned to Paris, Goudsmit had shirts made from the beautiful white material for all the Alsos members and most of their families back home.

In February and March 1945, Alsos's ranks began to expand as scientists were flown from the United Stated to Europe to join the mission in anticipation of the impending march into Germany. These first few precious weeks on German soil would present the greatest opportunity yet for the Alsos Mission to locate and confiscate German wartime scientific materials. This included the German nuclear program, of course. But the potential massive influx of scientific intelligence spanned diverse fields of research, and the backgrounds of the incoming scientists reflected this wide range. To support the new scientists, both support staff and infantry were added to Pash's command. On March 6, in recognition of the mission's success and growing stature, Pash was promoted to full colonel.

By late March, Pash and a handful of scientists and infantry met up once again with T-Force command. The conglomeration of intelligence units then made their way as a group to the city of Ludwigshafen on the west bank of the Rhine. A narrow bridge over the river connects Ludwigshafen to Mannheim on the other side, and it was here that T-Force along with Pash and the Alsos team would cross into the German heartland. For two days, the Alsos jeeps sat on the bank of the river and watched as the seemingly unending column of American machinery, tanks, armored vehicles, and jeeps thundered their way across the low bridge.

On the second day of the crossing, March 28, Pash spotted a small gap in the line of Allied tanks, and two of Alsos's jeeps, carrying Pash, Fiebig,

Beatson, and four others, seized the opportunity and darted in. Crossing the river and landing on the eastern shore, they immediately turned south and sped toward their first target in mainland German territory just a few miles down the road.

The city of Heidelberg and the old university there had long attracted some of the best European scientific minds. During the war, the city had become the site of the Kaiser Wilhelm Institute for Medical Research, where Walther Bothe, the esteemed nuclear scientist and experimental physicist, was conducting experiments using Germany's only working cyclotron. Capturing Bothe and his machine was Alsos's next top priority.

The road to Heidelberg was clear and Pash and his men made good time as they headed south. Compared to many of their other targets, the city was largely undamaged by the fighting; all the large buildings had been marked on their roofs with a bomb-deterrent red cross or red square. The city itself straddles the Neckar River, and all the bridges connecting the two halves had been demolished, leaving a single pontoon bridge as the only path to the south. Luckily for Pash and his men, who arrived in Heidelberg just after dark, their primary target in the city was at the university, located on the northern bank. Rather than risk waiting until morning, they decided to move directly to the institute labs. The laboratory manager was startled to see American troops arrive at his doorstep—the Alsos team were the first Americans to enter this area of the city—and he quickly handed over his keys. Pash's men worked fast, and by 10:00 PM they had Walther Bothe and a handful of other prominent scientists in custody and had begun searching the labs.

Early the next morning, Pash took a jeep and two men and set out in pursuit of another of their Heidelberg targets, Carl Krauch, who worked as the top official at the chemical firm IG Farben. Krauch's home was purportedly in the forested hills surrounding the city. When Pash and his men arrived in the area, they found that Krauch had left behind a rather magnificent house, built into the hillside, providing a dramatic sweeping view above the tree line of the valley and city below. The sizable home contained numerous bedrooms and seventeen additional bunks were found in the basement. This would become Alsos's forward command post throughout the rest of their time in Europe.

"Alsos Heidelberg Main." The house of a former IG Farben official, used by Pash and Alsos as a base throughout the rest of their time in Europe. *NARA II, RG 165, Box 161, Alsos Mission Album*

A second Heidelberg base, which would be named Alsos Rear, was also established closer to the city. In keeping with their penchant for spectacular accommodations, three mansions, all situated along Heidelberg's famous Philosophers' Walk, were secured by Alsos for this purpose. This roadway was named after the habit of university professors to walk there in moments of serious conversation or contemplation. It follows the forested top line of the hills above the Neckar River, overlooking the medieval castle ruins on the opposite bank. Rather than cross the bridge to get to the army mess, which was declared by one of the Alsos members to be "filthy and . . . disgustingly unhygienic," Alsos took control of a small beer garden just down the street from their new post. The restaurant's owner was happy to cook for the mission in exchange for the army food rations that they brought with them.

With targets and accommodations secured, the rest of the Alsos team began to make their way into Heidelberg, and their investigative work began in earnest. Goudsmit arrived in town a few days after Pash. His first task, questioning the newly captured Walther Bothe, made him very nervous—this was to be Goudsmit's first meeting with an enemy scientist whom he had known personally before the war.

Shuffling past the guard posted at the door, Goudsmit entered the small office where Bothe was being held. Hearing someone enter, the old scientist looked over. It took Bothe a moment to place the man standing in the doorway wearing the signature green of the US Army, but when he realized who it was his face lit up. A rather slight man, Bothe smiled at Goudsmit from underneath an impressive mustache as the pair warmly shook hands—a serious break of protocol when dealing with enemy informants, though Goudsmit couldn't have cared less. Bothe was thrilled to see his old friend and fellow physicist, not least because his attempts to adequately explain his work to the American soldiers who were holding him captive had only led to more confusion.

Never a supporter of the Nazi regime, Bothe had been removed from his post in the physics department at the University of Heidelberg when it was taken under Nazi control and had been continuing his work at the nearby Kaiser Wilhelm Institute for Medical Research. He took Goudsmit on a tour of his labs at the institute, proudly showing off the cyclotron. Goudsmit was again struck by just how far behind the German program was compared to the American effort. Bothe's cyclotron was the only working instrument in Germany; the United States had at least twenty, not to mention Lawrence's behemoth at Berkeley. Goudsmit was also surprised to see how little effect the war had had on the direction and scope of Bothe's work. While Bothe had certainly been busy, the German scientist's research was largely aimed at pure physics, rather than at harnessing physics for deadly effect.

The conversation then turned toward the information Goudsmit and Alsos were really after. When asked about uranium research, the old man, who had been jovial and pleasant, suddenly grew agitated. While Bothe was never closely tied to the Nazi Party, he was a fiercely loyal and nationalistic German, and he reprimanded Goudsmit for asking him to divulge information that he had sworn to keep secret. Prior to the Allies' arrival, he had even gone so far as to destroy all relevant documentation, notes, and data from his uranium work, a fact that stunned Goudsmit, who couldn't fathom destroying such hard-won results. Bothe would remain tight lipped about his work until the day that the war in Europe was officially over, May 8, 1945, when he finally submitted a full report of everything he knew about the German nuclear program—but by that point, his report was of little importance to Alsos or the Manhattan Project scientists.

As in the other cities that they had investigated, Alsos scientists made their way around scientific laboratories and offices in Heidelberg, gathering equipment and documentation wherever possible. But Goudsmit and his team didn't see much point in spending days painstakingly combing through these findings; they would only confirm what they had already learned. Alsos didn't need more clues, they needed to find the actual nuclear labs.

The Alsos team knew from the letters they had found in Strasbourg that Germany's nuclear research program had been split in half. An army physicist named Kurt Diebner had been conducting experiments out of Gottow, a military base south of Berlin. Alsos also knew that this lab had been evacuated in late 1944, and Diebner, along with his materials and research, had moved to the small town of Stadtilm in the central Thuringia region of Germany. The other nuclear research program headed by physics legend Werner Heisenberg had been moved from the Kaiser Wilhelm Institute for Physics in Berlin to a remote town in the Black Forest called Haigerloch. For the moment, both targets remained out of reach.

In February 1945 Roosevelt, Churchill, and Stalin had met at the Yalta Conference in Crimea to begin planning for the end of the war. The territory of Germany was divided into planned zones of occupation. As Germany fell, each of the major Allied powers would be given control over different portions of the country. While initially this division remained between the United States, Britain, and the USSR, Roosevelt and Churchill persuaded the Soviet leader to concede that the French should be allowed their own zone of occupation as well, provided that this area did not come out of the portion previously allocated to the Soviets. As the Allies entered Germany, France was projected to take control over two noncontiguous areas of western Germany. This created a problem for Alsos. All their remaining targets now sat in areas that were to be under French occupation, and General Groves didn't like or trust the French with nuclear information any more than he did the Soviets.

15 | DIEBNER'S LAB

Stadtilm, Germany
April 1945

AGITATED AS ALWAYS by the slow progress of the military impeding his progress toward his next objective, rather than wait for Allied progress, Pash began to formulate plans to access the remaining Alsos targets ahead of the front. They would try and go after Heisenberg's lab in southwestern Germany first. This area was still held by German forces but was about to be invaded by encroaching French troops. Concern about looting and lingering suspicions about Joliot-Curie's ties to the Soviets prompted panic-filled missives from Washington to Pash stressing the importance of getting to Heisenberg's lab before anyone else did.

To access this region as soon as possible, plans were made for an airborne operation. The Thirteenth Airborne Division would mount a paratrooper assault, dropping behind enemy lines and securing the area. Then, under heavy air protection, two Alsos scientists would be brought in to make whatever assessments they could and seize any available material. Two C-46 transport aircraft would land a few hours later to evacuate the Alsos team and captured German scientists, and return them to Alsos Rear at Heidelberg. The airborne division would then remain on the ground and begin working their way toward Berlin. Pash, ever the hands-on commander, planned to join the paratroopers on their plunge into the Black Forest. Goudsmit, on the other hand, made it clear that he thought the whole idea was utterly harebrained, saying he did not consider the minuscule German atomic research program "worth even the sprained ankle of a single Allied soldier." While Pash named his plan Operation Effective, Goudsmit insisted on referring to it as Operation Humbug.

With a little luck, both Goudsmit and sanity prevailed, much to Pash's disappointment. Early in the morning of April 10, the day the operation was supposed to get underway, Goudsmit was at Versailles, giving a briefing on the findings from Heidelberg to the US commanders at Supreme Headquarters Allied Expeditionary Force, when he was pulled aside and informed that General Patton was about to enter the Thuringia region, where Kurt Diebner's lab was located. While Goudsmit believed that the mission was most likely to find what they were looking for at Heisenberg's lab in the south, this second nuclear experiment program, run by the German army physicists, was likely to contain valuable intelligence and reactor supplies. They could not risk leaving it exposed. Goudsmit managed to get Pash on the phone just moments before he was set to leave for the airdrop operation. Heisenberg's lab would have to wait. Within hours, Operation Effective/Humbug was scrapped and Alsos began making plans to move into Stadtilm.

Goudsmit caught a ride back east with David Griggs, a geophysicist and pilot who was consulting for the air force and was in the rarefied ranks of civilians with access to military planes. When Goudsmit arrived outside of Stadtilm, "an army ordinance car was waiting for us, and we followed the troops. Then as soon as that little town fell a few hours later we were in there. It was a harmless village really."

Intelligence indicated that the uranium pile that Kurt Diebner had been working on was located in an old schoolhouse near the center of town. Goudsmit and the Alsos team expected to find a sizable operation when they arrived at the site—maybe not on the scale of the Manhattan Project's Oak Ridge, but impressive nonetheless—this was the site of the German army's own nuclear program, after all.

What they found when Pash's jeeps rolled into the courtyard of the dirty and run-down one-story concrete school building was quite the opposite. Inside, on the ground floor, they found a modestly supplied laboratory. Goudsmit noted the immense discrepancy between the assembled handful of small instruments and pieces of equipment and the massive facilities and mammoth devices in use in the US atomic program.

In the basement of the building, they found a dark cellar that appeared to have been at least partly a natural cave. Inside the basement, several dozen

Germans were huddled together. They initially claimed to be refugees trying to seek shelter from the fighting they thought was coming their way, but it didn't take long for Carl Fiebig to determine that these were actually the laboratory's scientists and their families. The basement room was entirely bare, aside from a deep pit standing empty in the center of the floor.

As Fiebig talked with the Germans, Pash suddenly recalled seeing something odd outside the building as they had hastily made their way in. As he went back up the stairs, one of the German scientists was escorted behind him as he walked out the door and around to the side of the building. There he found, stacked together in a loose jumble, several large white blocks of paraffin. Paraffin, Pash knew from his many conversations with his physicist teammates, has several possible uses in atomic research.

As Pash drew closer to the paraffin pyramid, he noticed something else. Scattered around haphazardly in the grass were several small pitch-black blocks.

"What is this black stuff here?" Pash asked, gesturing at one of the blocks. Fiebig translated the question to the German scientist.

"Oh, that is nothing but coal," the man replied hastily.

Pash bent over and picked one of the blocks up. Looking up at the German scientist and glancing over at Goudsmit, Pash smirked.

"It feels awfully heavy for coal."

The handful of pressed uranium oxide blocks that had been left behind on the lawn were not nearly enough for conducting experiments, much less building a serious attempt at a nuclear reactor, which the pit in the cellar floor suggested was the goal at this location. The rest of the uranium, then, was missing.

Also missing were the two top scientists who had been working there. The Stadtilm site had been the location of Diebner's experiments. But as the Alsos team came to discover, it had also served as the offices of Walther Gerlach, the physicist who had ostensibly been put in charge of overseeing both Diebner's and Heisenberg's nuclear programs for the last year of the war. Neither Diebner nor Gerlach were anywhere to be found. Gerlach, the scientists huddled in the basement told Pash, had left the area some time before, but Diebner had been at the site only two days earlier. A gestapo truck unit had shown up suddenly and ordered Diebner and a few key men to pack up all the materials that they would require to continue their work and had hastily evacuated the scientist and his experiment. Rumor had it that they were headed east toward the Alpine Redoubt in Bavaria, where the remaining Nazi fanatics were said to be gathering to make a last stand.

The town of Stadtilm was gloomy and gray, but not too cold for early spring in Germany, which was fortunate as the house that the Alsos Mission stayed in during their investigations had no heat or electricity. Goudsmit was able to rig a few lights from the ceiling, using batteries from Diebner's lab, which allowed the group to play cards at night. A wide range of taxidermy wildlife trophies adorned the walls of the abandoned family home, and the men made a game of hiding a stuffed goose inside a different person's bedroll each evening. A few days later, as they were packing up to leave, Goudsmit suddenly noted, with some discomfort, how closely the rug on the main bedroom floor matched the Saint Bernard featured in several paintings that hung from the walls.

However, Alsos's time in Stadtilm had been well worth the slightly macabre accommodations. The documentation that was found in Gerlach's lab largely confirmed what they already knew, but the exhaustive records he had kept in his office now began to add layers of detail and certainty. Alsos now had access to names, dates, locations of laboratories, and most interesting, financial paperwork, from which they were able to surmise the approximate amount of money that Germany had spent on nuclear research. All these details "merely reflected the pitiful smallness of the whole enterprise," and the Alsos men noted with amusement that the operations of the mission had likely already cost the American taxpayers more than the Germans had spent on the entirety of their atomic program.

16 | OPERATION BIG

Black Forest, Germany
April 1945

WHEN HE RETURNED FROM STADTILM, Pash found some unexpected visitors waiting for him at the Alsos base in Heidelberg. Michael Perrin, a leading figure in the UK's own nuclear research program, Tube Alloys, had come to Germany, bringing with him Sir Charles Hambro, who had been placed in charge of acquiring raw nuclear material for England with the help of the United States. Colonel John Lansdale, who had made his way to Europe at the beginning of April in preparation for the scrapped Operation Effective, was also present.

Allied troops were moving into northern Germany, and it was thought that the lion's share of the uranium powder that had been brought into Germany from Belgium might be located at a fuel storage facility near the town of Staßfurt. This area was to be within the Russian zone of occupation as Germany fell. Haste was required if the material in fact existed and was to be kept out of Soviet hands. First thing the next morning, Pash, Lansdale, and the two Brits headed north with the Eighty-Third Infantry Division. Skirting around active fighting in the Harz mountains, they arrived in Staßfurt on April 17.

Accessing the inventories of the depot outside town, it was clear that the barrels of Belgian uranium had been stored at this site. The warehouse had taken a beating from Allied bombing; parts of the facility were flattened, and the remainder stood only tenuously. But the massive stock of powdered uranium ore was located, stored in the same type of stacked barrels they had found in Olen and Toulouse a few months prior.

The facility had been abandoned for some time, though, and "most of the barrels were either broken or rotten." Russian forces were quickly advancing toward the area, and the group frantically searched for a way to get the uranium ore out before they arrived. Scrambling, Lansdale contacted a nearby paper bag manufacturing plant. He was able to secure ten thousand of their largest and sturdiest bags, and within two days the repacking and evacuation of the uranium ore was underway.

A battalion of trucks was dispatched from Allied headquarters in France to move the material to Hanover, and a Royal Air Force transport carrying British POWs moved a portion of it to England. The rest was hauled by rail to the port city of Antwerp for shipment to the United States. In total, almost 1,000 tons was moved out of the area over two weeks.

This supply of crushed uranium ore was almost all the material that the Germans had taken out of Belgium—they had apparently used only a small fraction of their available supplies. Here was yet more proof of the limited scale of the German nuclear program. Groves noted in a memo that "the capture of this material, which was the bulk of uranium supplies available in Europe, would seem to remove definitely any possibility of the Germans making use of an atomic bomb in this war."

While Lansdale and Hambro managed the evacuation of the ore from Staßfurt, Pash returned to Heidelberg, where another of the Alsos Mission's windows for action was narrowing. French troops were making rapid progress into German territory to the south, and if Alsos was going to try to beat them to Heisenberg's lab, it was now or never.

The expedition south, code-named Operation Big, was going to be the grand finale of the Alsos Mission, and the excitement was palpable. Not only had General Danvers authorized the use of the 1269th Engineer Combat Battalion to be commanded by Colonel Pash for this sortie, but the company included several other prominent guests. Michael Perrin, the British nuclear scientist, was the only one among the group who had actually seen a real working nuclear reactor; he had been in Chicago on an official visit to the States during Fermi's experiments and had been allowed to observe. Sir Charles Hambro also accompanied Alsos south along with several other of their British Tube Alloys colleagues. The American contingent included several new scientists, official personnel, an official photographer named Michael "Mickey" Thurgood, and John Lansdale representing General Groves. General Harrison, the US Army chief of intelligence for the southern area, came along as well. The plan was for the military contingent, led by Pash, to move out first. They would clear the way and secure the town of Hechingen, where Heisenberg and his colleagues were believed to be located. The team would then make their way across the river into Haigerloch, to Heisenberg's lab, where his nuclear experiment was being built. Only after the areas were secure would the scientists, led by Augustine, move out.

On April 20 at 3:00 PM, a stream of armored vehicles and jeeps made their way out of Heidelberg. About two hours into their trek, it started to snow heavily, and Pash decided that it would be best to stop for the night. But just after midnight Pash was shaken awake and handed a report that said the French troops were making a sudden push into Stuttgart and would overtake Alsos's targets in a matter of days, maybe hours. The snow had stopped, so Pash woke up everyone in his company and the column began a dark, cold, mad dash south.

On April 22 the column reached the Black Forest and a command post was set up along the Neckar River. Rather than head toward Hechingen as initially planned, Pash changed focus: with limited time, his priority became capturing Heisenberg's lab. A small group of armored vehicles set out for Haigerloch. This tiny town nestled in a deep gulley carved by an S-bend in the Eyach River looked untouched by the war that had been raging outside the valley. The Royal Bridge served as the only entrance or exit to the village, and a single cobblestone road ran through the town and up the hillside to the opulent gilded Renaissance church perched on top of a cliff face, overlooking the collection of age-old houses below. As they crossed the bridge and entered the town, they passed through a flurry of bedsheets and white shirts held out on the end of broomsticks, waving from the windows on either side of the narrow street.

Locating Heisenberg's laboratory in the tiny town was an easy task. A boxlike concrete entrance had been constructed at the mouth of a small cave carved into the side of an eighty-foot cliff. The cave, which had once served as a *Bierkeller* (beer cellar), was situated directly underneath the church, high up on the hill above. The location was shrewd; the cave provided protection against arial observation, while the church provided protection against bombardment.

The cliff side and concrete bunker where Heisenberg's lab was located, Haigerloch, Germany. *Virginia Military Institute Archives, Colonel John Lansdale photo collection #0000089*

When Pash and his men arrived at the base of the cliff, they found the doors of the concrete entrance padlocked shut, but a paper attached to the doorframe indicated the manager's identity. Pash's men inquired about the manager's whereabouts at the local pub, and he was quickly located. When Pash demanded, through Carl Fiebig's translations, that the manager open the lock, the man made some vague attempts at dissuading their entrance, saying he did not have the correct keys. But Pash, so close to his final target, had no patience for games or desire to be slowed down and ordered his triggerman Beatson to shoot the lock off the door.

Moments later, dynamite charges did away with the concrete entrance in its entirety. Pash and his team had initially intended to blow up the whole cave but were stopped after the first blast by Monsignor Gulde, who was responsible for the church looming above. He came flying down the 149 steps that connected the church to what was once its cellar and threw himself, screaming wildly in German, between the Americans and the cave entrance. The blasts might compromise the stability of the ancient *Schlosskirche*, and no translation was required for the Americans to understand that this would not happen while the monsignor remained alive. Pash relented, and while the concrete entrance structure was now a pile of rubble scattered at the mouth of the cave, the interior of the makeshift laboratory remained unharmed.

In the center of the cave, Pash and his men found a deep circular pit, roughly six feet wide, that had been carved out of the stone floor. The inside of the pit was lined with a cylindrical layer of black bricks, and inside this wall sat a large metal drum. This central container was capped with a thick and heavy metal lid. But this pit, which had been built to house Heisenberg's "uranium machine," was empty.

Toward the back of the chamber sat a work area containing tools and workbenches, alongside two electrical control panels attached to the wall. Pash noted that the panels had each been given names, in what was clearly some kind of inside joke. The upper one was labeled PROFESSOR DIEBNER, while the lower was apparently called PROFESSOR GERLACH. Three large cylindrical containers were attached to the far back wall of the cave and connected by a series of pipes to the experimental pit in the center of the chamber. These drums, which now stood empty, had once held Heisenberg's supply of heavy water. In front of the water tanks sat a small blackboard on which someone had scrawled:

> *Die Ruhe sei dem Menschen heilig*
> *nur Verrückte haben es eilig*

Pash roughly translates the text thus: "Let rest be holy to mankind. Only crazy people are in a hurry."

Heisenberg's nuclear laboratory had been found, but still missing were the uranium, heavy water, and nuclear scientists. While the reactor had been built in Haigerloch, the German physicists had set up their offices in Hechingen, just a few miles down the road.

Leaving his own scientists, who had arrived at the scene, to gleefully explore their new find, Pash set out to find Heisenberg and his colleagues. Thanks to information that they had gathered many months ago in Strasbourg, Pash and his team already knew exactly where to go when they arrived in Hechingen: a wool mill on the edge of town. Within fifteen minutes of arrival, the makeshift offices of the evacuated physics labs and six other buildings were

"Doctors, officers EM and CIC agents all pitch in to dismantle this installation in the cave laboratory." *NARA II, RG 165, Box 161, Alsos Mission Album*

occupied by Allied soldiers. Within half an hour, Alsos had captured nearly all the physicists on their priority list, including Carl von Weizsäcker, whom they had been searching for since he had evaded them in Strasbourg. The only man missing was Werner Heisenberg, who they were told had left some days earlier, apparently traveling by bicycle, to be with his family at their summer home in the Alpine lake town of Urfeld.

While Heisenberg was nowhere to be found, his office had been left virtually untouched. When Pash walked into the abandoned room the first thing to catch his eye was a framed photograph of Heisenberg standing next to none other than Samuel Goudsmit. The photo had been taken during one of the summer physics sessions that Heisenberg had attended at the University of Michigan before the war. The circumstances of this photo and why it was there had to be quickly explained to a very alarmed General Harrison, and Goudsmit endured a lot of teasing about the photo in the weeks to come.

As their list of wanted scientists dwindled, Pash, along with Larry Brown, Beatson, and Fiebig quickly moved on toward the nearby hamlet of Tailfingen, where the offices of the Kaiser Wilhelm Institute for Chemistry had been relocated. This institute was headed by none other than Otto Hahn, who, with Lise Meitner, had discovered nuclear fission just over five years earlier. Though Tailfingen was located just a few miles down the road, the team found that moving through this region, more heavily fortified by German forces and not yet under Allied control, was extremely slow and precarious. The small convoy moved from village to village, "taking these towns more or less by surprise and by telephone."

When the Alsos team arrived at Hahn's makeshift lab in an old school building, they found Otto Hahn waiting for them with his bags packed. With him was the venerable old German physicist Max von Laue. Hahn insisted that, despite his early connection with fission, neither he nor Laue knew much about the work their colleagues had been conducting down the road.

However, Hahn did have a grim message to relay to Sam Goudsmit. Like Heisenberg, Hahn was an old friend of Goudsmit's from their university years. Following the German invasion of the Netherlands, both Hahn and Heisenberg had been trying to keep track of Goudsmit's elderly parents who were trapped in The Hague. After learning that they had been detained, Hahn and Heisenberg had sent letters to any Nazi official they thought might listen to try to secure the Goudsmits' release. But the letters had had little effect, and when he saw his old friend in Hechingen, Hahn was the first to tell Goudsmit that his parents had been murdered in the gas chambers.

As for the missing uranium and heavy water, the cohort of captured German physicists kept insisting that they could not speak on the matter without direct permission from Heisenberg, who was still, conveniently, missing. It was the opportunistic Carl von Weizsäcker, hoping to gain the favor of his new captors, who was eventually the first to crack. He told Pash that the data and other documentation from the lab had been carefully wrapped before being dropped into the cesspool of a nearby outhouse. These documents were fished out of the outdoor latrine by unlucky GIs and deposited below the window of the building in town where Goudsmit had set up an office. The rising smell alerted him to the package's arrival.

Inside the rancid bundle, Goudsmit found the data to prove what he had suspected for months: the Germans had not been successful in creating a functioning nuclear reactor. They hadn't even been close. In fact, Goudsmit reasoned, as he scribbled out the math on a scrap of paper, the data suggested that they would have needed a reactor that was at least 50 percent bigger, both in uranium and heavy water moderator to have a chance of achieving criticality. Goudsmit was struck by how small the sum total of Germany's nuclear program, now laid out in full before the Americans, really was. The entire program took up little more than a cave, and a few rooms in a textile factory and school building.

In addition to the location of the data, Weizsäcker also revealed where the actual reactor materials had been hidden. The uranium that had been part of the pile at Haigerloch had been buried on the edge of a nearby farmer's field, and the heavy water had been secreted in a nearby abandoned gristmill. Everyone was eager to be part of the recovery of this precious material and a large group rushed up the hill to the location Weizsäcker had indicated. In the mill they found three drums of heavy water stashed in a dark, cobweb-covered corner. These were rolled out and loaded onto a waiting truck.

The old water-driven gristmill near Haigerloch where three drums of heavy water were seized. *NARA II, RG 165, Box 161, Alsos Mission Album*

In a field high up in the hills, overlooking the town below, they searched for the German uranium, with the eagerness of pirates digging for buried treasure. A German technician from Haigerloch was made to do the actual digging, to protect against possible booby traps. Pash recalled the moment when, "excavating on the edge of the plowed field, we next struck a heavy wooden shield, the removal of which exposed nearly the entire core of the Nazi atomic pile in uranium ingots."

One by one, cubes of uranium metal were pulled out of the ground and stacked in the dirt. The cubes were uniform in size but varied slightly in appearance. Some had been roughcast, with voids and craters speckling their surface. Others had been more carefully machined, their faces smooth and dark. Still others had been created from sandwiching together layers of centimeter-thick square plates. The one thing that all had in common was the notches that were filed into their edges, which had held in place the aircraft cable from which they had been hung.

Cache of uranium elements found by the Alsos team buried in a plowed field. *Virginia Military Institute Archives, Colonel John Lansdale photo collection #0000089*

Alsos members pulling the buried German uranium out of the ground. *Virginia Military Institute Archives, Colonel John Lansdale photo collection #0000089*

Uranium metal is incredibly dense, and moving the material out of the field was difficult. Several soldiers attempted to pick up small boxes full of a few dozen cubes and were astonished to find that they could not lift them. A human chain was formed to move the cubes, one by one from the underground vault into the bed of an awaiting army truck.

Alsos now had everything. News quickly spread of the capture of the sum total of Germany's nuclear program, and positive proof that Germany was not a nuclear threat. A message was relayed up the ladder from Pash to Eisenhower to the Pentagon and finally to the White House and President Truman himself, now in only his second week in office. All the messages read the same: "Boris Pash has hit the jackpot."

Though they had achieved their primary goal, Pash's task was not yet complete. They had captured the laboratories and most of the German scientists, but still missing were the three men who had been primarily responsible for Germany's nuclear program: Heisenberg, Diebner, and Gerlach. All three were reported to be somewhere in the Nazi-occupied Bavarian Alps to the east. Diebner and Gerlach were rumored to be near Munich, while Heisenberg was in the lake town of Urfeld.

On the first day of May 1945, Pash and a small team made their way from Heidelberg southeast toward the Upper Bavarian Alps, meeting up with the Thirty-Sixth Infantry Division. The war was clearly in its final days, and there were signs of surrender hanging from every window.

Pash's march east was slowed when they came to the town of Kochel on May 2 and found that a bridge in their path had been destroyed. The town of Urfeld lay across the bridge and the only way around this roadblock was to hike through the snowy woods over the Kesselberg mountain. Army engineers were working furiously to repair the bridge, but Pash was, as always, in a hurry. He took a company of nineteen men, and they started the arduous hike into the Alps. The group traveled light, taking only what they would absolutely need and leaving their heavy packs behind. As they made their way up the mountain, snow halfway up their shins, they encountered several small groups of German soldiers who, upon seeing the Americans, immediately surrendered. Unable to take the prisoners with them, Fiebig instead cut each man's waistband, so he had to use his hands to hold up his trousers, and these men were sent down the hill with instructions to surrender to the next American they saw.

The unit finally topped the wooded ridge of Kesselberg at 4:40 PM to a spectacular view of the large clear lake surrounded on all sides by forested snow-covered hills. Around that lake was situated the idyllic mountain town of Urfeld.

The descent down the mountain was somewhat easier than the ascent had been. An inn by the road became their command post when they arrived on the outskirts of town. Early the next morning, Pash and a couple of his men headed to Heisenberg's house, located about halfway up the mountains on the opposite side of the basin. As they approached the cozy wooden cabin, they found Werner Heisenberg, Germany's preeminent physicist, waiting for them on the veranda.

News of Hitler's death had begun to spread, and the remaining resistance in Germany crumbled. Another small Alsos team had traveled to Munich, where, on May 1, 1945, they found Walther Gerlach's wife at home alone. She accompanied the Americans to the university, where her husband was working in his basement offices in the physics laboratory. Gerlach was taken into custody. The head of Germany's nuclear program quickly gave up the location of the team's last target, Kurt Diebner, and on May 2, a small team was dispatched to Schöngeising, a small town outside of Munich, to collect him.

With all three major targets now secure, the Alsos team and their captives returned to Heidelberg on May 4, with weeks of interrogations and reporting ahead of them. Just four days later, on May 8, Germany officially surrendered and the last remaining anxieties about a nuclear Nazi Germany finally dissipated.

The lab where the B-VIII experiment was conducted has now been turned into a museum. Visitors who make their way to the isolated town of Haigerloch, somehow still untouched by time, can enter the cavern of the Atomkeller Museum (after paying four euros) through two huge wooden doors. Inside, the space is small but surprisingly densely packed with explanatory panels (all in German), miniature dioramas, and full-size object displays. An aluminum scale model of the reactor core, as it would have appeared with the experiment fully assembled, is suspended over the deep circular pit in the center of the room. Against the back wall, ominously lit by blue floodlights, stand the three drums that once held the precious heavy water, with a replica of the chalk note containing the musings on serenity and hurry leaning against them.

While the cubes hanging from the model B-VIII core are not uranium, real uranium artifacts sit in a display case to one side. The Atomkeller Museum houses both a cube of uranium metal, and a one-centimeter-thick square plate of uranium, which would have been lashed together in sets of five to approximate cube dimensions. The sign posted above the display says that the cube was transferred to the museum by Germany's department of environmental protection. But as the sign also notes, while of the correct approximate weight and dimensions, the cube lacks the characteristic black oxide layer, or pitting,

Author photo

that coats other known cubes from Germany's World War II stock. Also lacking are the notches, typically seen carved in the edges of the cubes, allowing them to be safely suspended in place by the aircraft cable connecting them together. Instead of these features, this cube sports a clean solid surface with distinct machining marks and a mysterious set of numerical markings. Whether this cube is genuine, or an approximation of the historical objects made by a more modern machinist, is a question only detailed expert scientific analysis could ever hope to address.

PART II

THE REACTOR
HITLER TRIED
TO BUILD

German cube reactor experiment: B-VIII. *AIP
Emilio Segrè Visual Archives, Goudsmit Collection*

17 | MODERN PHYSICS

IN THE EARLY DAYS of the twentieth century, physicists around the world found themselves singularly focused on one objective: unlocking the secrets of the atom. While the idea of an atom, the smallest component unit of matter, had originated centuries before in ancient Greece, it wasn't until the late nineteenth and early twentieth centuries that a more sophisticated and experimentally derived understanding of the atom, its behavior, characteristics, and most important, component pieces, began to evolve.

The first model of the atom emerged at the University of Cambridge in 1897 when J. J. Thomson discovered that even though an atom as a whole has a neutral electric charge, it contains small, extremely light, negatively charged particles, which Thomson named electrons. Thomson knew that for an atom to be electrically neutral, the negative change of the electrons had to be balanced out by something with a positive charge. Based on his experiments, Thomson developed the first model of an atom that assumed a uniform positively charged atomic body, with small negatively charged electrons scattered throughout.*

When Ernest Rutherford came from New Zealand to Manchester, England, in 1907, he set out to test Thomson's version of the atomic model. Earlier in his career, he had discovered alpha particles—a particle emitted from certain radioactive materials like radium. While he was still unsure of what these particles were, he knew they were relatively large (subatomically speaking) and

---

* The so-called plum pudding model.

that when they struck a surface of zinc sulfide, a tiny bight dot of light could be observed. In Rutherford's laboratory, his assistant Hans Geiger (for whom the Geiger counter would be named) began aiming beams of these particles at thin sheets of different materials, and observed the resulting spread of the dots of light on a zinc sulfide plate on the other side as the alpha particles passed through the foils. Most of the particles traveled straight through, lighting up the plate behind the film, but Rutherford and Geiger began to wonder what was happening to any particles that did not make it to the other side. Rutherford assigned his student, Ernest Marsden, the task of finding out.

In what would become one of the seminal experiments in the history of physics, Marsden bombarded an incredibly thin foil of gold with alpha particles. Instead of measuring the particles that passed through the foil, he looked for any particles that might be bouncing back toward the beam source. Based on Thomson's model, all the particles should have easily made their way through the foil to the detector on the other side. But Marsden observed that about one in every eight thousand alpha particles did not reach the other side of the foil but was instead deflected back in the general direction of the alpha particle source. This phenomenon could only be explained if the particle was hitting something within the atom it was passing through—something large, but localized—a nucleus. A new model of the atom was proposed based on this discovery, of a cloud of negative electrons surrounding a central positive nucleus.

In 1913 Niels Bohr took this model one step further when he suggested that electrons do not just surround the nucleus but orbit it in predetermined distances, or shells. In this model, the orbits of electrons correspond to well-defined energy levels that rise with increasing distance from the nucleus. The electrons in the outermost orbital shell therefore have the highest energy level.

At this point, study of the atom began to diverge. The study of electrons—their energy and movement within and between atoms—would develop into quantum mechanics, while the study of the nucleus—its components and the forces holding it together (or breaking it apart)—would become the realm of nuclear physics.

At the center of many of the significant developments in modern physics in the early part of the twentieth century was the Kaiser Wilhelm Society in Berlin. Founded in 1911 to promote the development of the natural sciences in Germany, the Kaiser Wilhelm Society comprised several smaller "institutes," each focused on a specific area of scientific inquiry. These institutes existed outside the traditional academic university structure and were intended to function as world-class research centers where scientists were given the space and funding to devote themselves to their work without the burden and bother of teaching university students. By the time Hitler rose to power in 1933, the Kaiser Wilhelm Society was arguably the most significant center of science in the world.

This was particularly true for the Kaiser Wilhelm Institute for Physics, which rose to global prominence as the home of Albert Einstein. Einstein, a German Jew, was brought to the institute by prominent German physicist Max Planck in 1914. Two years later he produced a paper describing a new concept that would revolutionize the physics world: general relativity. This theory would come to redefine the lens through which physicists observe the forces at play in the world around them, and it opened up a literal universe of scientific study. By 1917 Einstein had been offered a permanent position at the Kaiser Wilhelm Institute for Physics, and in 1921 he was awarded a Nobel Prize for this work.

But Einstein was a revolutionary in more than one sense. In addition to his massive contributions to modern physics he was also an outspoken internationalist, a Zionist, and loud opponent of Europe's pervasive anti-Semitic sentiments. He was also able to claim a title that few scientists before or after him have ever been able to achieve: he was a celebrity. His was a household name, making his outspoken opposition to the growing Nazi Party all the more problematic in their eyes. Einstein just so happened to be in the United States at the moment Hitler's National Socialist Party gained control of Germany. Rather than return home, he renounced his German citizenship and took a position at the Institute for Advanced Study in Princeton, New Jersey, that he would hold for the rest of his life.

While Einstein's insights cracked open a new world of possibilities, it was another young, but brilliant German student who made this new frontier his own. The immense intellect of Werner Heisenberg, who was born in 1901 in Wurzburg Germany, was clear from a young age. At nineteen, he had enrolled in the physics

"Werner Heisenberg on the occasion of his Scott Lecture at Cavendish Laboratory, Cambridge, England." *AIP Emilio Segrè Visual Archives, Bainbridge Collection. Werner Heisenberg B7*

program at the University of Munich, working under Arnold Sommerfeld, and within just a few months of study, young Heisenberg had already gained a firm enough grasp of the material to begin correcting the work of his fifty-two-year-old mentor. A year later, Heisenberg published his first paper, and after following this first publication with three more, Heisenberg suddenly found himself at the forefront of the developing field of quantum mechanics. This new field, and the flurry of work and insights it introduced, turned many of the underlying conceptions of traditional classical physics on their head.

In 1927 Heisenberg introduced the idea for which he is best known and still bears his name today: the Heisenberg uncertainty principle. The next year, his work had earned him a professorship at the University of Leipzig, making him the youngest full professor in Germany. And in 1933, a month before his thirty-second birthday, his contributions to physics earned him a Nobel Prize.

The other defining force in Heisenberg's life was his marriage to Elisabeth Schumacher. The pair met in January 1937 at the home of the owners of a publishing house in Leipzig, where the twenty-two-year-old Elisabeth had just started working as an apprentice. When she met the thirty-five-year-old Nobel laureate and professor, he was playing piano. "Family lore has it," their eldest daughter later wrote, "that Werner won Elisabeth's heart with Beethoven's 'Largo con espressione' from the piano trio Op.1, No. 2." Within two weeks they were engaged, and by April, they were married.

Throughout their marriage, Elisabeth would become Heisenberg's closest confidante. The pair stayed in near-constant contact, and their letters, which are filled with talk of music, the small details of life, and a great deal of love, offer a window into the life and mind of the great physicist during the course of the looming war.

Heisenberg's early and astounding success in physics was not the result of accident or fate but of a deep-seated drive to excel. In her memoirs, Elisabeth wrote of her husband that his "family of pedagogues" had bred a tireless pursuit of academic excellence. His father had been a professor of classics, and Heisenberg had been introduced to the writings of the Greek philosophers from a young age. As a young man, he discovered a Greek inscription over the doorway of a school that translated roughly to "Always strive to be the best at whatever you do," and he "decided to make this his guiding motto."

In some instances, Heisenberg's adherence to this personal creed resulted in a fierce competitiveness. Elisabeth describes a trip to Shanghai during which Heisenberg was trounced in a ping-pong match. He responded to this failure by playing the game nearly nonstop on the voyage home, until he was able to beat anyone on board who was willing to play him. Family games with his children, Elisabeth noted, presented a similar challenge.

Heisenberg was also fiercely nationalistic. As a young man he had participated in the Pfadfinder, a branch of the German youth movement. As an adult and as a scientist, while he frequently expounded on the virtues of a global community of physics, these statements were accompanied by the thinly veiled subtext of Heisenberg's belief that among this global community, German science and German intellectualism were—and ought to remain—at the top of the heap. A test of his belief in German superiority and his loyalty to an idealized vision of German science would come sooner than he might have imagined. Only weeks before he was awarded his Nobel Prize in Stockholm, Hitler and his Nazi Party had taken power in Berlin.

———————

As Niels Bohr and Heisenberg were busily working to describe the movements and interactions of electrons, work was also proceeding on the study of the atomic nucleus. After scientists ascertained that a small dense nucleus lay at the center of every atom, it was posited that this nucleus might itself be made of other subatomic components. These positively charged components, which balance out the electrons' negative charge, were named protons by Ernest Rutherford.

However, the concept of a tightly packed mass of positive particles presented a problem: the protons, with their like charges, should not clump together but instead should repel one another. Therefore, Rutherford assumed, there must be another component within the atomic structure that exists inside the nucleus along with the protons helping to hold the nucleus together. He reasoned that this mysterious third type of subatomic particle must have a neutral charge, and so he named them neutrons. It would take another eleven years for James Chadwick to experimentally confirm their existence in 1932.

A clear picture of the stable atomic nucleus was beginning to emerge. The identity of an atom as a given element on the periodic table is determined by the number of positive protons at its core, and, in a neutrally charged atom, the number of protons in the nucleus will equal the number of electrons surrounding it. But the number of neutrally changed neutrons in a nucleus was found to vary slightly, resulting in different isotopes for each element. Because it is the presence of the neutrons within a nucleus that prevent the protons from repelling each other and breaking the atom apart, it followed that changes in the number of neutrons within the nucleus of a given element can affect its overall stability.

In some elements, the forces at play between the protons and a given number of neutrons results in an unstable atom. These unstable nuclei will spontaneously break apart into smaller, more stable components and emit radiation in the from of energy, alpha particles, and most important, lone neutrons. This phenomenon of radioactivity had been observed first by Henri Becquerel in 1896, and then later studied and defined by Marie and Pierre Curie.

Work by the Curies' daughter, Irène, and her husband Frédéric Joliot-Curie in 1933 took this work further, by showing that bombarding the nuclei of certain stable elements with alpha particles could cause them to emit their own radiation. This meant that radioactivity was not just a spontaneous process—it could be induced. In Italy in 1934, Enrico Fermi took the next step, by systemically bombarding foils of nearly every element on the periodic table with neutrons instead of the larger and positively charged alpha particles that Rutherford had used. This experiment, when repeated by Otto Hahn and Fritz Strassmann on uranium foil, would result in the discovery of nuclear fission, an extreme form of radioactive decay, in 1938.

Before the beginning of World War II, the community of scientists working on the study of atoms, both electrons and nuclei, was a small and tight-knit group. Conferences were more like reunions, as old classmates met up with past mentors to discuss new developments in their work.

All of this changed in the summer of 1939. Heisenberg and his young and sardonic protégé Carl von Weizsäcker traveled to the United States to attend the Summer Symposia on Theoretical Physics at the University of Michigan. The annual event was held on campus, and as in past years, Heisenberg spent the week at the home of his old friend and classmate, Samuel Goudsmit.

These gatherings had long been a high point in the academic year of the participants, but this year, the tenor of the meeting was noticeably more subdued as the encroaching war hung heavily over every conversation. This would likely be the last time, they all knew without saying, that physicists from all over the world would be gathered together; battle lines would soon be drawn between them.

Many of Heisenberg's American colleagues hoped that when those lines were drawn, Heisenberg would be included in their own ranks. Heisenberg had been grappling with the decision of whether to emigrate from Germany for some time. Elisabeth describes his struggle in her memoir: "He wrote [to Bohr] he almost wished that he had no other choice than to emigrate, that, as with so many of his friends, fate would decide the question for him." But no such sign from on high appeared, and Heisenberg was left to weigh the options on his own. In a letter to his wife, Heisenberg wrote:

I have been asked by others a few times under what conditions I would permanently come here. The two aspects, light and shadow, are so immensely clear. I am treated fabulously in every way. I would have ten times as many bright students as where we are. It would probably also affect the results in my work positively. But we are just not at home here. The children would speak English and grow up in an atmosphere that is foreign to us. That would not be nice at all, and so we are just staying put.

Newly married, Heisenberg was intently aware of the implications of moving his young family to America. Ultimately, he could not let go of the ambition to raise his children in the idealized Germany of his mind, regardless of how far from reality that vision was becoming. Nor could he let go of the notion that he himself might have some vital role to play in rebuilding Germany into the intellectual center he believed it could and should be after the coming war. At the meeting that summer, when asked by his friends if he would consider staying in the States, his answer was a refrain that he would often repeat throughout the war and for many years after: "Germany needs me."

It would seem that in some ways, Heisenberg either genuinely or willfully did not understand the depth of the divisions that the coming war would create. Elisabeth writes, "At the time, he still firmly believed in the 'world community of physicists' he spoke about so often, and he was totally unaware that he was excluding himself from this community through his decision to stay in Germany."

At a party one evening, as the conference in Michigan drew to a close, Edoardo Amaldi, the same Italian physicist whom Boris Pash would spend a great deal of time and resources frantically trying to apprehend in Italy in just a few years' time, looked around the room at all the assembled men. He leaned over to a friend and remarked, "See Fermi, see Heisenberg, sitting in a corner. Everyone in this room expects a big war and the two of them will lead fission work on each side, but nobody says." By this time, Fermi had already begun a series of small experiments on uranium back at his home base at Columbia University, but he volunteered little information on the matter to Heisenberg during their time together that summer.

At the beginning of August, Heisenberg and Weizsäcker were among a small list of passengers bound for Germany on one of the last ships back to

Europe. They were returning, as Goudsmit put it, with "the conviction that we were sound asleep in the United States, that we were doing practically nothing on the uranium problem—just ordinary routine, purely academic nuclear physics. It was a good state of mind for our enemies to have. It lasted them and comforted them until the end of the war."

18 | JEWISH PHYSICS

THE GERMANY TO WHICH Heisenberg and Weizsäcker were returning at the end of the summer in 1939 was not the gleaming intellectual utopia in Heisenberg's mind but a Germany undeniably changed by a regime that had brought the country to the very brink of an all-out global war. Nazi rule had radically changed nearly every facet of German life. Academia, where Jews and other so-called non-Aryans made up a large proportion of the academic body, had certainly not been spared.

On January 30, 1933, the same year that Heisenberg was awarded the Nobel Prize in Stockholm, Adolf Hitler was installed as the new German chancellor. The new Nazi government moved swiftly, and on April 7 under the Law for the Restoration of the Career Civil Service, Jews, Communists, and other "non-Aryan" individuals were formally prohibited from working as teachers, professors, or judges. A similar law was soon passed pertaining to lawyers, doctors, musicians, and several other professions. Jews currently working in these positions were expected to resign, or, the new Nazi government made clear, they would be forcibly removed.

Within a few months, Jewish professors began to be dismissed from university posts across Germany. The Kaiser Wilhelm Society fired the majority of its Jewish scientists in the spring of 1933 when they received notice that, though the society was not a university, the new rules pertained to them as well.

In September 1935 the Nuremberg Race Laws expanded the Nazis' definition of who was classified as Jewish, lengthening the roster of academics slated for removal. By April 1936, nearly 40 percent of Germany's academic body had been removed from university posts.

Many of these exiled intellectuals made their way west, often to the United States, where this sudden influx of talent strengthened and revitalized American labs and universities. Einstein was among the first of these European scientists to move to the United States, though throughout the 1930s and into the early 1940s, dozens of the biggest names in physics would follow, including Leo Szilard, Edward Teller, Hans Bethe, Niels Bohr, and Enrico Fermi. Some of these men were fleeing their own persecution; others, like Fermi, left their labs and careers in Europe to protect their families. Still others left in solidarity, unwilling to stay and work in a country that was headed down such a dark road.

Through the early twentieth century, American science had largely been viewed on the global stage as somewhat provincial, lagging behind its European counterparts. As mathematicians, chemists, and physicists began to pour into American universities, the status quo in global science swiftly began to shift in the United States' favor.

But physicists like Fermi, Szilard, and Bohr brought more with them than just their reputations and prestige. They brought their fledgling but visionary understanding of atomic physics and nuclear fission. In a twist of historic irony, many of the refugees who had been removed from their posts and forced to flee their homes by the Nazis would soon play critical roles in the success of the US nuclear weapons program.

Conspicuously missing from the ranks of German physicists who left Germany was Werner Heisenberg. For Heisenberg, whose fierce nationalism was inextricably tied to the superiority of German science, leaving was unthinkable. A self-professed "apolitical" scientist, Heisenberg clung, however naively, to the notion that the pursuits of modern physics were entirely separate from and transcended the political abominations brought by the Nazis.

What Heisenberg either did not yet see, or not yet acknowledge, was that the assault on German physics and academics in general by the Nazis extended beyond the political to the ideological. On May 10, 1933, the German Student Union had led a ritualistic burning in Berlin, and other cities in Germany and Austria, of books that they considered contaminated by "Jewish intellectualism." Among the books that were thrown into these bonfires amid speeches and "fire oath" incantations were the works of Albert Einstein.

The wholesale rejection of any and all ideas or developments made by non-Aryans, which had begun in the streets, quickly made its way into university physics lecture halls, where the burgeoning principles of modern physics were subordinated to anti-Semitic ideals. Two men in particular spearheaded the assault against physics in the universities: Phillip Lenard and Johannes Stark.

For Lenard, who had received a Nobel Prize in 1905 and who had been a fervent supporter of the National Socialists long before Hitler came to power, Einstein's brand of modern physics, with its complex thought experiments and confounding mathematical underpinnings, was baffling. Unable to wrap his own mind around these new ideas, Lenard instead drew the only conclusion that his inflated ego would allow—that relativity and quantum mechanics were foolishness. Instead, Lenard insisted that the only true physics was "German physics," or better still, "Aryan physics," which amounted to little more than restatements of traditional classical physics descriptions. Modern physics, often personified by Einstein himself and labeled as "Jewish physics," was, according to Lenard, the product of the "decadent Jewish spirit" and therefore could be wholly discredited as "alien"—a stance that the Nazi regime was all too eager to embrace.

The relationship between modern academic science and the ideology of the Nazi regime was tenuous at best. As Samuel Goudsmit put it, "Before the war the Nazis had voiced their prejudices against science and scientists loud enough for the whole world to hear. The cold logic of science did not fit very well into the mystic cult of blood and soil." Throughout the war, scientists remaining in Germany were flabbergasted to find themselves suddenly competing for funding and support with ideologues offering fantasy and magical thinking presented as serious scientific work.

These fanciful ideas would come to define Nazi science. Many of the most absurd concepts were drawn from the Schutzstaffel (or SS), the paramilitary organization that became the epicenter of Nazi ideology. From the mythology of Teutonic ancestry and the alphabet soup of stolen and bastardized runic symbols, the SS developed many of its own pseudoscientific theories. For example, they held the inexplicable belief that they referred to as "*Welteislehre* or world ice theory" that at the center of the earth, as well as all the planets and stars, is a core of solid ice.

However, the most grotesque and insidious presentation of "Nazi science" came in the form of human experimentation on concentration camp prisoners.

From amputations performed without anesthesia, to exposure to diseases and toxic chemicals, to experiments involving extreme pressure and extreme cold, Jews and other captives of the Nazi regime were subjected to unspeakable torture throughout the war, as Nazi scientists pursued their malignant theories.

From his seat at the University of Leipzig, Heisenberg watched in horror as the ideological perversions of physics offered by Lenard and Stark took hold, and the principles of modern physics, the crown jewel of German scientific superiority, which Heisenberg himself had played a large part in developing, began disappearing from classrooms and lecture halls across the country.

As a leading voice in German physics, Heisenberg felt that it was his duty to take on the "spectre of national hubris manifesting itself so clearly in 'aryan physics.'" Heisenberg continued to teach relativity in his classrooms and gave a series of lectures defending the viewpoint of "true physics" as he saw it as well as defending the physicists, his friends and colleagues, who had developed it. In February 1936, he even published an article in the *Völkischer Beobachter*, a typically pro-Nazi newspaper, attempting to calmly and reasonably explain to a public audience the importance of not abandoning true physics. But as Elisabeth astutely observed, "Heisenberg had underestimated his opponent."

At the time of his marriage to Elisabeth in April 1937, Heisenberg was finishing up what was intended to be his final semester of teaching at the University of Leipzig. He had been offered an appointment at his alma mater, the University of Munich, taking the place of his retiring doctoral supervisor Arnold Sommerfeld.

When the newlyweds arrived in Munich that July, a few weeks later than expected due to illness, Heisenberg called the university chancellor to let him know of his arrival and ask for more information regarding his new appointment. Heisenberg was surprised by the terseness of the chancellor's voice on the other end of the line.

"Have you already seen *Das Schwarze Korps*?" the chancellor asked, referring to the pseudonews publication of Hitler's SS.

Heisenberg, who did not make a habit of reading the propaganda paper, had no clue what the chancellor was referring to. He ran down to the nearest shop and bought a copy of the July 15 issue. Inside, he was stunned to see the headline WHITE JEWS IN THE SCIENCES. The article, penned by none other than Johannes Stark, contained a full-throated personal and ideological attack on Heisenberg himself, referring to Heisenberg and his theoretical-physics colleagues as "Einstein disciples."

While Heisenberg was singled out in this article, the attacks were aimed more broadly at modern physics, linking it to Judaism and declaring it decidedly non-German. The article read: "Heisenberg is only one example among several others. All of them are puppets of Jewry in German intellectual life and must disappear just as the Jews themselves." In a few paragraphs, Stark had pitted Heisenberg, his work, and all of modern physics squarely and explicitly in opposition to the Nazi regime.

Heisenberg was at a loss for what to do. This article was surely going to cost him his new position at the University of Munich. But beyond that, Heisenberg feared, it might cost him his reputation as an expert and authority of the highest caliber. Suddenly, Heisenberg found himself relegated to a similar fate as his Jewish colleagues. Shunned by the Nazi establishment, he now faced a choice of leaving Germany or staying and risking his career and perhaps his life if he continued to speak about and teach modern physics.

Heisenberg once again weighed his options, and once again found himself unable to leave his beloved Germany to its fate. "Heisenberg was personally much too involved in his fight against 'German physics,'" Elisabeth would later write, and he believed that "his departure would have meant a fateful victory for this aberrant teaching."

Heisenberg's old friend Samuel Goudsmit casts Heisenberg's decision to stay in a somewhat different light: "Heisenberg [was] a man of ideals, but ideals distorted by extreme nationalism and a fanatical belief in his own mission for Germany." Heisenberg still hoped that the grip of the Nazi regime would be short-lived. "One day," he had said to Goudsmit in the days before the war, "the Hitler regime will collapse and that is when people like myself will have to step in."

Unwilling to leave Germany, and unable to abandon the teachings of modern physics, Heisenberg instead took a huge gamble in his quixotic quest to defend physics from the ravages of Nazi rule.

On July 21, 1937, Heisenberg decided that the only course of action was to plead his case directly to Heinrich Himmler, the leader of the SS and the man ultimately in charge of what was printed in their propaganda mills. Heisenberg wrote a letter in which he argued for the inherent neutrality of science and scientific principles and demanded that his name be cleared and that he be protected from further such attacks as had appeared in *Das Schwarze Korps*.

Getting the letter to the high-ranking Nazi was no simple matter. To ensure that Himmler received and read his plea, Heisenberg turned to his mother for help. Heisenberg's mother, Annie Wecklein, was, as it turned out, a distant relation of Himmler's own mother. She arranged to meet with her cousin and transferred Heisenberg's letter over tea.

In writing this letter, Heisenberg knew he was taking a massive risk. If the SS leader disagreed with, or worse, was insulted by Heisenberg's appeal, it was well within Himmler's growing power in the Nazi regime to permanently silence the physicist.

He would have to wait months to receive a reply, during which time Heisenberg became the target of a full-blown investigation by the SS. He was repeatedly called in to Berlin's infamous Prinz-Albrecht-Strasse prison, where he was subjected to extensive questioning by the notorious author of some of the greatest horrors of the Holocaust, Reinhard Heydrich. Certain that their conversations were being monitored in their home, Heisenberg and Elisabeth took to taking walks outside in order to speak openly.

It took a year for Himmler to make up his mind about what to do with Heisenberg. Finally, in a letter to Heydrich sent in July 1938, Himmler says that after reading Heisenberg's message and giving the matter some thought, he "believe[s] that Heisenberg is decent." Also, he notes as an aside, Germany really "cannot afford to lose [Heisenberg] or have him killed, since he is relatively young and can bring up the next generation."

Himmler wrote a letter to Heisenberg officially absolving the physicist of the accusations that had been leveled against him and assured him that he would be protected from attacks like these in the future. But a clear warning had been added in the letter's postscript: "PS. I do find it appropriate, though, that in the future you separate clearly for your students acknowledgment of scientific research results from the scientist's personal and political views."

Heisenberg had dodged a bullet, and he returned to his teaching post at Leipzig thinking that perhaps he had won some sort of moral victory against the Nazis for the truth. In reality, with his absolution from Himmler he had entered into a Faustian bargain with the Third Reich. Heisenberg had been given a reprieve, but it had come with a cost: he was now indebted to the Nazi regime and expected to fall in line. He would be allowed to continue teaching and working on modern physics, but in return, by staying in Germany, and later, by serving as a sort of scientific ambassador to occupied countries during the war, Heisenberg had surrendered his fame and stature to legitimize the regime he claimed to detest.

Heisenberg was far from the only German scientist who chose to stay in Germany rather than flee westward. Otto Hahn, Walther Bothe, Max von Laue, and many others decided to stay and attempt to continue teaching and conducting research as best they could. But as these scientists worked without their Jewish colleagues and with the suspicious eyes of the Third Reich peering over their shoulders, physics in Germany became just a shadow of its former glory.

19 | THE URANIUM CLUB

JUST LIKE WITH THE UNITED STATES' Manhattan Project, it was a letter that served as the initial spark igniting Germany's own nuclear weapons program. Recognizing the potential of Meitner and Hahn's discovery, on April 24, 1939, just weeks after the publication of the article in *Nature*, Paul Harteck, a professor at the University of Hamburg and explosives consultant for the German Army, along with his assistant Wilhelm Groth, wrote a letter to Erich Schumann, the physics professor turned head of weapons research for Germany's Army Ordinance (Heereswaffenmat). This letter described the potential possibilities that the newly discovered process of atomic fission might hold for the German military in the coming war— perhaps even as a terrible new weapon. In part, the letter read: "We take the liberty of calling to your attention the newest developments in nuclear physics, which, in our opinion, will probably make it possible to produce an explosive many orders of magnitude more powerful than the conventional ones. . . . That country which first makes use of it has an unsurpassable advantage over the others."

Unbeknownst to Harteck and Schumann, several weeks earlier, two other scientists, Georg Joos and Wilhelm Hanle, had contacted Abraham Esau, the head of the physics department of the Reich Research Council (or the RFR, after its German title, Reichsforschungsrat). In 1937, the RFR had been organized to oversee and centralize all scientific research being conducted in Germany under the oversight of the Ministry of Education. On April 29, in response to Joos and Hanle, the council organized a preliminary meeting to discuss the feasibility of a fission research program. They nicknamed their gathering the Uranverein, or the Uranium Club.

When Eric Schumann learned that a nuclear research project was already underway at the RFR, he moved quickly to bring the project under the Army Ordinance's control. The army assembled their own crude version of an atomic research program before issuing orders to Esau and the RFR to cease their own fission work. When Esau complained, he was told that the army had already been working on nuclear fission for years (despite the discovery being less than a year old) and that wasteful, duplicative work should be avoided. Esau, recognizing the futility of arguing the matter any further, relented, and a second version of the Uranium Club was officially organized under the Army Ordinance. To lead this "new" initiative, Schumann chose a little-known army physicist named Kurt Diebner.

In mid-September, the newly appointed Diebner and his assistant Erich Bagge convened a meeting in Berlin to discuss the practical possibilities of uranium wartime research. In attendance at this gathering were many of the biggest names in German physics and chemistry, including Otto Hahn, whose experimentation had led to the discovery of fission earlier that year; Paul Harteck, whose letter had, in part, spawned this whole endeavor; and Walther Bothe, who, as one of the oldest men in the room, had served as mentor to many of the men seated around the table.

To this list of attendees, Bagge had also extended an invitation, almost as an afterthought, to his own mentor and adviser, Werner Heisenberg. While arguably the leading physicist remaining in Germany, Heisenberg, a devout theoretician seated at the table full of experimentalists, was the odd man out. Most of the research and work that had made him famous was based in mathematics and thought experiments, not in constructing experimental setups out of concrete and steel, or in the mundanity of collecting actual data. But Heisenberg's international stature and fame made him welcome among his colleagues at the meeting.

Rather than start from scratch in developing their nuclear program, at the beginning of October 1939 the Army Ordinance took formal control of the Kaiser Wilhelm Institute for Physics (KWIP) in Berlin, placing its administration and research under their influence, and positioning the institute to serve as the primary site for the nuclear research efforts. The KWIP's sitting director, Peter Debye, was given an ultimatum: either formally renounce his

Kurt Diebner. *Wikimedia Commons*

Dutch citizenship or leave his post. He chose the latter and hastily made his way to the United States, taking a position at Cornell University. In his place, Schumann installed Diebner as acting director of the KWIP.

Born in central Germany in 1905, Diebner had studied physics at Martin Luther University of Halle-Wittenberg and received his doctorate in 1932. After working for three years as a teaching assistant to nuclear physicist Gerhard Hoffmann, he had taken a job as a researcher at the Reich Physical and Technical Institute, where he served as an adviser to the Reich Ministry of Defense and the Army Ordinance Office.

While Diebner's background and education had given him a reasonable understanding of nuclear physics as well as expertise in explosives, he was a total outsider among the academic elite of the KWIP. Without any real connection to the various seats of power and influence in academic physics and lacking any awards or credentials that came even close to the Nobelian gold that typified past institute directors, Diebner was immediately sized up by the scientists of the KWIP as their intellectual and academic inferior. He was often referred to in letters and whispered conversations by these same men, both during and after the war, with a kind of academic slur, which, though relatively innocuous to laymen's ears, was intended to cut the man it was aimed at to the quick: a second-rater. It was under this cloud of derision that Diebner began the task of trying to guide the Uranium Club members toward their first goal of building a self-sustaining nuclear reactor.

20 | HOW TO BUILD A NUCLEAR REACTOR

WHILE POPULARIZED VERSIONS of nuclear reactor control rooms, with innumerable blinking lights and buzzing alarms, make it seem like a nuclear reactor is unfathomably complex, in reality, a reactor really only has four component parts.

All reactors require fuel. In the early days of nuclear fission, before the development of isotope separation, this fuel came in the form of natural abundance uranium, that is, uranium with the same ratio of the two major isotopes (^{235}U and ^{238}U) as occurs in nature. As there was still no way to increase, or enrich, the percentage of the fissile ^{235}U in their fuel elements, scientists on both sides of the Atlantic required large amounts of uranium (on the order of a couple of tons) in order for a reactor experiment to contain a high enough concentration of ^{235}U. Several different forms of this material could be used. The majority of Fermi's uranium in the first American reactor, CP-1, was pressed uranium oxide powder, though uranium converted into its metallic form was also included in the pile.

A nuclear reactor also requires a moderating agent. Moderators are a unique class of material that slow the movement of any neutrons that they come into contact with, without absorbing or stopping them. When a uranium atom fissions, the neutrons that are released are moving too quickly and, at that speed, can be captured and absorbed by atoms of ^{238}U. A moderator material is placed between uranium fuel elements in order to slow the movement of flying neutrons enough that they are able to reach an atom of ^{235}U and precipitate another fission event. In order for this to work, the size of the fuel elements and the distance between them (the length of the path through the moderator) have to be precisely controlled. Carbon, hydrogen, and deuterium (a heavier isotope of hydrogen) were identified as effective moderators, and were used in various formats (e.g., graphite, water, heavy water, and paraffin) in early reactor experiments.

The third component of a functioning reactor is a neutron source that emits free neutrons in order to get a fission chain reaction started. And the last component is a series of control rods, made of materials that are very efficient at absorbing neutrons. It is the control rods that, as their name suggests, actually control the reactor.

In a reactor that is not "on," the control rods are fully inserted into the uranium core, absorbing most of the neutrons emitted by both the neutron source and by natural fissions that occasionally occur in the uranium fuel elements. "Turning on" a nuclear reactor is as simple as slowly removing the control rods from inside the pile. As the neutron absorbers are removed, some of the neutrons from the neutron source, instead of quickly being absorbed, are able to find their way through the moderator to a neighboring uranium fuel element and then produce a fission event.

The two or three newly freed neutrons that are a result of that fission are again able to travel through the moderator to another uranium element, creating yet another fission—a chain reaction. If left unchecked, the number of fission events would exponentially increase as three neutrons become 9, become 27, 81, 243, and on and on. However, it is possible to control this rise in the neutron population using the control rods. If the control rods are still partially inserted into the core, some proportion of these neutrons will still be absorbed, preventing the number of neutrons being generated from climbing too high, too fast.

A self-sustaining nuclear reactor has been achieved when the number of free neutrons in the system at any time remains constant; that is, the number of neutrons causing fissions is equal to the number being generated by those fissions minus the number lost or absorbed. This state, once reached, is called criticality. Building a reactor that could reach and maintain criticality was the first task facing Germany's Uranium Club in their pursuit of nuclear power.

Following their initial meeting in Berlin, Heisenberg, the preeminent theoretician, returned to his lab at the University of Leipzig where he began to work through the mathematical framework underpinning the construction of a critical reactor. In December 1939 he produced the first of two reports based on this work. It was indeed possible, he surmised, to build a self-sustaining fission chain reaction using uranium. Heisenberg extrapolated that if a large enough mass of the minor fissile isotope of uranium, ^{235}U, could be obtained, "the entire radiation energy of all available uranium atoms would be set free all at once," resulting in an explosion many times larger than any weapon that had ever been made before.

But, while Heisenberg's initial reports seemed promising for the progress of the German nuclear program, in his calculations and assumptions, Heisenberg had made several critical mistakes. The first was his calculation of how much uranium would be required. The critical mass, or the amount of uranium required to create a self-sustaining chain reaction or an explosive instantaneous reaction, is in part related to the distance a neutron must travel between fission events. Heisenberg assumed that this distance was much larger than it actually is, and as a result his initial calculations of a critical mass of ^{235}U were on the order of several tons—far larger than the actual size required.*

Whether out of fear of repercussions, respect for the revered scientist, or a widespread lack of understanding of the problem at hand, none of Heisenberg's colleagues corrected his mistake. This miscalculation, which was taken as fact, presented the German scientists with a problem. Separating out several tons of ^{235}U from several hundred tons of natural uranium would be difficult and extremely expensive. And even if enough material was somehow obtained, the sheer weight of the resulting weapon would largely rule out the possibility of an aerial delivery.

As a result of these impediments, the German scientists began to pull back from the notion of using nuclear fission to build a weapon. Instead, they turned their attention, and their rhetoric about their research, toward using this new technology as a source of energy. Their primary goal became building a functioning nuclear reactor—or a "uranium machine," as they called it—in the hope of providing energy to power Germany's war industry.

Heisenberg's second mistake, which he outlined in a second report in February 1940, was concluding that graphite, pure solid carbon bricks, would

---

* In reality, the critical mass of ^{235}U is only about forty-six kilograms.

not work as a moderating material. Faulty analysis from Walther Bothe, who was working in Heidelberg in 1941, further cemented this error. Bothe's results suggested that graphite, one of the two possible moderating materials available, absorbed too many neutrons to be feasible. What no one would realize until it was too late was that the graphite being considered for these determinations was not clean enough. Too much of the natural boron contamination found in most sources of graphite was skewing their results. Scientists in the United States had also encountered this problem but had found that properly purified graphite does indeed work well as a moderator, and was the material chosen by Fermi in the construction of CP-1 just under two years later.

With graphite off the table, the German scientists were forced to turn to heavy water, which was costly and difficult to produce, as their primary moderating material. Throughout the war, the primary source of heavy water for Germany's nuclear research was in occupied Norway. In 1906 the Norsk Hydro company had dammed the waters of the Måna in Telemark, a region in the south of the Scandinavian peninsula. The waters from the dam were used to fuel the Vemork power station—at the time the largest hydroelectric plant in the world, which provided power to the electrolysis facility next door. Here the electricity generated by the dam was used to split molecules of water, also taken from the Måna, into hydrogen and oxygen gas for use in the production of ammonium fertilizer. An incidental result of this electrolysis process is the production of heavy water.

An atom of the lightest element in nature, hydrogen, typically consists of a single electron orbiting a nucleus with only one proton, and no neutrons. But in one out of every 6,420 hydrogen atoms, a neutron will find its way into the nucleus. This slightly larger isotope of hydrogen is called deuterium.* When deuterium replaces the two hydrogen atoms in a molecule of water (H_2O) it becomes heavy water (D_2O).

---

* Much rarer, one out of a quintillion (10^{18}) hydrogen atoms has a second neutron, bringing the total number of protons and neutrons in the nucleus to three, hence this isotope's name—tritium. Tritium is an unstable isotope and is the central component of hydrogen bombs.

The existence of heavy water was discovered in 1931 by Harold Urey at the National Bureau of Standards in the United States. Soon after this discovery, the chemists at the Norsk plant realized that this by-product of their production might be a valuable resource. Rather than continuing to discard the deuterated water, Norsk Hydro decided to move forward with plans to use it. A facility was constructed in the basement of the plant, and a series of new electrolysis cells were put in place. Each step in a seven-part series further concentrated and purified the liquid, until finally, drop by drop, nearly pure heavy water was collected.

When Germany invaded Norway in 1940, the German chemical conglomerate IG Farben took over the production operations in Telemark, producing both valuable fertilizer for German agriculture and heavy water for use in the nuclear research program.

Initially, obtaining adequate supplies of uranium also presented a challenge. Most of the world's uranium supply in the early twentieth century was being mined from the Belgian Congo by Union Minière. In 1936 the Belgian government declared that the country would remain neutral in any coming conflict, effectively preventing the sale of its massive uranium supplies to Germany. While a lesser grade of uranium was being mined from the Joachimsthal region of Czechoslovakia, then under German control, obtaining enough purified uranium from this source alone was going to be costly and slow.

With the German occupation of Belgium in May 1940 came unfettered access to hundreds of barrels containing thousands of tons of uranium ore powder stored at the Union Minière warehouse near Brussels. Auergesellschaft, a rare-metal company from Berlin, took over the contract for refining and processing much of this material, opening a new plant in Oranienburg. In June, the company ordered the first shipment of sixty tons of uranium ore to be sent to Germany for processing and use in the nuclear research program. German scientists now had access to all the raw uranium they might need.

21 | EARLY GERMAN EXPERIMENTS

RATHER THAN ESTABLISH a single centralized location from which to conduct all their work on uranium, the German scientists largely returned to their own labs and universities after their initial meeting in Berlin. Diebner, now the head of the Kaiser Wilhelm Institute for Physics located in the suburbs just south of Berlin, began directing the work of the institute's scientists. Bothe returned west to the University of Heidelberg and his cyclotron lab at the Kaiser Wilhelm Institute for Medical Research, and Heisenberg began his theoretical analyses while he returned to his lab and his teaching schedule at the University of Leipzig. Each had been assigned a piece of the project on which to focus his work.

The very first actual nuclear reactor experiment undertaken in Germany was constructed not by Diebner, nor by the preeminent Heisenberg, but by Paul Harteck, one of the authors of the letter that had brought nuclear fission to the German army's attention. Harteck, working out of his laboratory at the University of Hamburg, began building his experiment in early 1940, well before any large amounts of heavy water or uranium were readily available.

By this time, graphite had already been effectively ruled out as a moderator material by Heisenberg, and no significant supplies of heavy water were yet available. Harteck, who would come up with several of the German nuclear program's more creative solutions throughout the war, decided to try using another possible moderator: carbon-rich dry ice. Dry ice is frozen carbon dioxide. It does not suffer from the same impurity issues that graphite does, and Harteck believed the lower temperature at which dry ice must be kept to prevent it from evaporating might just make a fission chain reaction more likely.

For his experiment, Harteck obtained ten tons of the super-cooled carbon dioxide from the Leipzig-based chemical company Ammoniakwerk Merseburg GmbH. To transport the material from the factory to Harteck's lab in Hamburg, Diebner had to ask the War Department to lend the project a special refrigerated railcar.

Harteck's next challenge then became obtaining enough uranium to run his experiments. In January 1940 the first batch of purified uranium oxide powder was produced by Auergesellschaft. But rather than being delivered to a central location, this relatively small amount of uranium had instead been distributed among each of the several groups that were working independently at their own labs and institutions across Germany. Every researcher involved in the program was anxious to study the properties of uranium oxide in the hopes of perhaps being the first to build a "uranium burner," and no one was eager to surrender their small stash of black powder for the benefit of their competitors.

While more uranium was on its way, Harteck was working on a tight schedule. The dry ice would be needed in the summer for the storage of food and would have to be returned in a few months' time. Eventually, after much pleading and discussion, Diebner was able to convince several of the project scientists to temporarily relinquish their uranium supplies so that Harteck could build his experimental pile. Only Heisenberg notably refused to give up his stash. In the end, Harteck amassed about two hundred kilograms of uranium oxide powder and constructed a pile in which the black powder was poured into cube-shaped wells that had been cut out of stacked sheets of dry ice.

The frozen pile, which weighed about ten tons and stood seven feet high, was constructed in a freezing-cold meat locker to keep the dry ice from sublimating away. The extreme cold, and the potential for carbon dioxide poisoning as the dry ice slowly turned to gas and filled the room, made for miserable working conditions. And in the end, all the effort, cajoling, and misery did not pay off—there was far too little uranium in the pile, and the experiment did not show any appreciable results. The dry ice and uranium were returned to their respective points of origin in the late spring.

With the promise of more material on its way, the construction of the next reactor experiments began at two other locations in the second half of 1940.

In Leipzig, Heisenberg, working with Robert Döpel, began putting together a series of layered experimental piles. Each of the experiments in this series was designated L (for Leipzig) followed by its number in the sequence.

The L-I experiment consisted of four spherical layers—a sort of nuclear onion. A round central aluminum casing, filled with water, sat at the center of this experiment. The water core was surrounded by a spherical layer, a few centimeters thick, of uranium oxide powder held in place by another larger aluminum casing. Yet another larger metal shell held in place a second layer of water, and a final casing created an outer layer of uranium oxide. Results from this experiment, once assembled, were largely inconclusive, as were the results of a second version of the pile, L-II, which replaced the normal water with 150 kilograms of heavy water.

For Heisenberg, a theoretician in both training and in practice, working with Döpel on constructing these experiments was novel and intimidating territory. In July 1940 he wrote to his wife about learning some of the more practical aspects of experiment building from Robert Döpel's wife, Klara, a lawyer turned physicist working with her husband on the experiments in Leipzig: "So overall I am really busy, have stood this morning in the laboratory in a white coat and learned from Mrs. Döpel how to make metal tubes airtight. . . . I enjoy the opportunity to learn the basics in experimental physics."

Work was also beginning on a reactor experiment in Berlin at the KWIP under Kurt Diebner's direction. In July 1940 Karl Wirtz began overseeing the construction of a new dedicated laboratory space for the program. In the center of this small building sat a two-meter-deep cylindrical pit into which a reactor vessel, shaped like a drum and measuring 1.4 meters deep and wide, could be lowered and raised by a crane on the ceiling above. The unassuming facility was coined the Virus House, and signs were posted to discourage curiosity from passersby. By October 1940, the KWIP scientists were ready to begin construction of their nuclear reactor experiments.

Unlike Heisenberg, who chose to address the problem of reactor construction from a theoretical standpoint, Diebner was an extremely practical experimentalist—closer to what today might be considered an engineer than a true physicist. While Heisenberg favored the simpler construction and more elegant mathematics of a layered reactor design, when Diebner approached the same problem of optimal reactor geometry, he came to a very different conclusion. Instead of a two-dimensional layered structure, Diebner's design called for a three-dimensional lattice. Small chunks, preferably spheres (but cubes would do), of uranium would be distributed inside a larger body of moderator material at precisely defined distances from one another. Diebner's design plan, from the size of the uranium pieces to the distance between them, was eerily similar to the final structure of Fermi's successful nuclear reactor experiment, which was still two years away from fruition.

But Diebner would have to wait to test his three-dimensional design. The scientists working under him, particularly Karl Wirtz and Carl von Weizsäcker, did not like or trust the "second-rate" outsider who had been brought in to direct them. Instead of following Diebner's lead, they came up with excuse after excuse for Heisenberg to travel from Leipzig to Berlin by train every few weeks to "consult" with them on their work. Heisenberg insisted that his layered deigns were superior, and no one in the German scientific ranks was prepared to contradict him. As a result, the first reactor experiments constructed at the new facility in Berlin used Heisenberg's preferred layered design.

While the symmetry of the spherical layers of the Leipzig series of experiments simplified the necessary calculations, the design made it difficult to vary the number and thickness of the various layers. For the first "B" (for Berlin) experimental pile, a horizontal, layered construction was used. The central reactor drum was filled like a cake, with layers of uranium oxide powder separated by layers of paraffin wax. The drum was lowered into the pit, and the surrounding space was filled with water. When a neutron source was lowered into the pile, no neutron multiplication was observed. A second experiment soon followed that used two layered piles inside the central cylinder; still, no results were observed.

Back in Leipzig, between work on the reactor and his teaching load at the university, along with his increasingly frequent trips to Berlin, Heisenberg was getting weary of the frantic pace of wartime research. To make matters worse, he now frequently found himself alone. When Heisenberg had "resolved to subject [his family] to the coming catastrophe" by staying in Germany, he had purchased a small cabin in the tiny town of Urfeld on the shore of the Walchensee in the Bavarian Alps. In an attempt, "if possible, to spare the children the chaos of the bombing raids" and to keep his young family out of harm's way, Heisenberg had moved Elisabeth, their five young children, and his mother, Annie, to the miniscule lake town.

While the remoteness of the town shielded Heisenberg's family from the bombings, life there was difficult. Writing about her time in Urfeld, Elisabeth Heisenberg recalled that "the soil was rocky and barren, and the little that did grow was predictably eaten by the deer. In addition, the farmers harbored an implacable, distrustful stinginess toward us as outsiders. In fact, we had serious difficulties, and we waged a grim battle against hunger and sickness." Heisenberg tried to help his young family as best he could, sending crates of jars, canned goods, and fruit to the "eagle's nest" in the Alps, but most of these packages either never arrived, or showed up weeks later rotten, pillaged, or crushed.

Lonely and unhappy back in Leipzig, Heisenberg frequently wrote to his wife describing, in vague terms, the day-to-day business of his work. In a letter sent in the summer of 1940 he bemoaned his fate at the center of both experiments: "This life of an industrial tycoon is no fun for me at all; I need tranquility of the soul to be reasonably productive." But no tranquility awaited him.

In Heisenberg's lab in Leipzig, Robert and Klara Döpel were continuing work on the spherical layered reactor experiments. For their next trial, a change was made to the format of the uranium they received from Auergesellschaft. Early iterations of both the L and B experiment series, as well as Harteck's dry ice experiment, made use of uranium oxide powder, which was relatively easy to manufacture and to work with: it could be poured into hollow containers to create fuel elements of various shapes with masses that could be precisely controlled by scooping a bit out, or sprinkling in a little more, to arrive at the desired weight. But uranium oxide powder is less dense and less pure than metallic uranium, a difference they thought might be hindering their results.

In 1940 Auergesellschaft's uranium refinement plant at Oranienburg did not have the technical means for converting the oxide powder to metal. To solve this problem, laboratory manager Nikolaus Riehl reached out to the head of Auergesellschaft's parent company, Deutsche Gold- und Silber-Scheideanstalt (or Degussa), out of Frankfurt. Degussa had previously developed a method for the production of thorium metal from oxide and found that the process, which used a hefty supply of calcium, could be used in the production of uranium metal. Degussa soon began making metallic uranium to supply Germany's reactor program. The metal Degussa produced was not in the form of bars

German pile sphere. *AIP Emilio Segrè Visual Archives, Goudsmit Collection*

or pellets but was instead a granular metallic power. It was hoped that this product could provide the physicists with the advantages of uranium metal while still allowing it to be used in the reactor experiments in much the same way as the uranium oxide powder. On May 1, 1942, the Degussa company delivered 3.5 tons of uranium metal powder to the German nuclear researchers.

The third Leipzig pile (L-III) was assembled in a seventy-five-centimeter-diameter spherical aluminum shell, containing two layers each of heavy water and powdered uranium metal. The experiment came to an abrupt end, however, when the uranium powder suddenly burst into flames, destroying the sphere along with the entire supply of uranium and heavy water inside.

Like rusting iron, uranium metal oxidizes readily when it comes into contact with air. For larger uranium components, this oxidation and the resulting layer of black powder that forms on the surface can be controlled through the application of coatings. But for metallic uranium in powder form, this oxidation is difficult to control and can be extremely dangerous. In contact with air, the metal powder can oxidize so quickly that the reaction can produce enough heat for the whole supply of uranium to spontaneously catch fire, a property known as pyrophoricity.

Uranium fires burn hot, capable of reaching a peak temperature of 1,500 degrees Celsius—hotter than the temperature needed to melt glass. They are also difficult to extinguish; water is ineffective, and smothering with graphite or sand is slow.

Rather than rethink their design, Heisenberg and Döpel doubled down. Another larger experimental uranium-and-heavy-water onion was constructed in the summer of 1942. More metallic uranium powder was obtained from Degussa and layered with 140 kilograms of heavy water in an eighty-centimeter sphere. The whole apparatus was sunk into a pit filled with water.

The increase in size paid off when a neutron detector at the surface of the pool began to measure slightly more neutrons being emitted from the sphere than were being dumped into it by the radium/beryllium neutron source at its center. While they were nowhere near a self-sustaining critical chain reaction, fission events were clearly occurring within the experimental pile.

But Heisenberg's elation at their first real success was short-lived. Days later, while Heisenberg was out teaching a seminar, Döpel noticed bubbles beginning to rise in the experimental pool. The gas in the bubbles was identified as hydrogen, a concerning result as it indicated that the surrounding water might have penetrated the outer aluminum shell and was reacting with the outermost layer of uranium powder.

Döpel, not fully understanding what was happening in the pool below him, decided that the best course of action was to lift the sphere out of the water to try to salvage the materials for future use. This was a mistake: the reaction occurring between the water and the layer of uranium oxide had created a negative pressure inside the outermost shell. When the sphere was lifted out of the pool, air rushed in through the fissures in the aluminum casing. The sudden influx of oxygen once again ignited the uranium powder, sending a long flame shooting out of the side of the sphere into the room. Panicking, Döpel doused the sphere with water, temporarily minimizing the flames, but he was not sure what to do next. He sent an assistant to fetch Heisenberg from his lecture.

A breathless Heisenberg rushed into the lab to find a terrified Döpel and a red-hot sphere smoldering as it swung from a crane in the center of the room. Worried that they would once again lose their entire supply of uranium and heavy water, Heisenberg decided that the best thing to do would be to lower the sphere back into the pool of water to try to cool it off.

This was another mistake. When the cool water rushed into the expanded fissures in the aluminum casing and hit the extremely hot uranium, it vaporized. The water in the tank began to ominously bubble as the metal sphere squeaked and popped and started to swell. Heisenberg and Döpel rushed out of the room just moments before the gases in the spheres burst their confines. The aluminum casing was torn open, and hot uranium powder was flung into the air, where it immediately caught fire, consuming the room. The fire brigade arrived minutes later and spent the next two days trying to fully extinguish the flames. All the uranium and most of the heavy water was lost.

An indignant Döpel wrote a letter to Auergesellschaft after this second catastrophe saying that he had not been adequately warned of the flammable properties of uranium metal. The company wrote back saying that "the tendency of uranium to react with air and water was well-known to every chemist, and that if in spite of everything physicists were ignorant of this, they ought to have looked at a circular sent out by Degussa a year ago, warning of these very consequences." In response to the disaster, a decision was made to use solid uranium metal elements in all future experiments, and though it had resulted in the first neutron multiplication measured in Germany, the L-IV pile marked the end of the Leipzig series of experiments.

22 | COPENHAGEN

BY THE TIME THE UNITED STATES finally entered the war in December 1941, the scientific communities on both sides of the conflict had been almost entirely cut off from one another for two years. While the American scientists could not be sure about the pace and scope of the nuclear work being conducted by their German counterparts, there was more than enough cause for alarm. Not only had fission been discovered in Germany but German universities—not to mention the KWIP—still boasted some of the best scientific minds in the world. And while the Manhattan Project had only formally begun working toward the goal of a nuclear weapon at the beginning of 1942, there was no reason to believe that work in Germany had been similarly delayed, and it was assumed that the German scientists were working with a two-year head start.

For the Americans, getting a glimpse into the German nuclear research program was difficult. Occasionally, the contents of a letter sent by a scientist inside Germany to colleagues in Switzerland or other neutral territories would provide a little insight, but it was impossible for Groves and his team to judge whether these reports could be trusted.

Conferences and technical meetings also provided an avenue for the Americans to gather information. Scientists working in Germany, like most other German citizens, were largely prohibited from leaving the country. An exception was made, however, for some of the most prominent German scientists to attend conferences and technical meetings that were still being held throughout Europe. Groves and his team regularly sent spies to these meetings to attempt to glean any information they could.

Much to his annoyance, Heisenberg was a common fixture at these events. What he had not at first fully comprehended was that, in exchange for his absolution by Himmler, following his run-in with Stark and the SS, Heisenberg had agreed to become a sort of scientific ambassador for the Third Reich. He suddenly found himself forcibly dispatched on a regular basis to occupied and neutral countries throughout central Europe to present his work at conferences and symposia. Often, these meetings were staged at so-called German cultural institutes, which had been erected in these territories to spread pro-Nazi propaganda under the thinly veiled guise of sharing culture and academics. The presence at these institute events of reputable speakers like the globally renowned Heisenberg lent legitimacy to these events and their messages.

In September 1941 one such conference was organized at a German cultural institute in Copenhagen, and Heisenberg, with his ever-present disciple Carl von Weizsäcker in tow, was asked to attend and present lectures. Unlike many of his trips, Heisenberg had actually been looking forward to this one, as it gave him the opportunity to visit his old friend and mentor Niels Bohr. During his visit, Heisenberg gave a lecture at the Niels Bohr Institute where he himself had once been a student, and that evening he had dinner with Bohr and his wife. After dinner, the student and his mentor took a walk. The conversation that took place between the two scientists that night has long been a subject of intense speculation.

After the war, Bohr and Heisenberg would each present polar-opposite views of what transpired that evening. Bohr, struck by Heisenberg's seemingly blind nationalism and willingness to make excuses for the atrocities that were being carried out by the Nazis, understood Heisenberg's thinly veiled references to his atomic program as boasts about the weapon that they were working to create. Heisenberg's wife, Elisabeth, believed that Bohr's interpretation of her husband's motives stemmed from Heisenberg's decision to stay in Germany as Bohr "knew that golden bridges had been built for Heisenberg's emigration to America, and that he had turned them down."

For his part, Heisenberg would later insist that his intentions were quite the opposite of how they were received: he had been attempting to elicit advice from his mentor about whether scientists could or should ethically play a role in war. Elisabeth again offers her own interpretation of Heisenberg's side of events, writing that her husband had become isolated and lonely in Germany, cut off from the larger physics world and unable to engage with other scientists on topics of greater interest to him than the nuclear work. Bohr had in many ways been a father figure to Heisenberg, and he had come to Copenhagen to seek both connection and absolution from his mentor.

Whatever the truth of the matter, after that evening, the pair would never meet as friends again. When Bohr escaped Denmark in 1943, narrowly avoiding an assassination plot, to join the Manhattan Project's work at Los Alamos, he presented Robert Oppenheimer with a sketch that Heisenberg had made during his visit. Drawn on the scrap of paper was a clear depiction of a reactor experiment with the layered design that Heisenberg had so doggedly been pursuing. The Americans took this sketch and Bohr's assessment of his student's motives as further proof that the German nuclear program was making rapid progress.

Heisenberg and Weizsäcker likewise left Copenhagen with their own impressions. During their trip, Weizsäcker had been able to acquire several of the more recent issues of the American journal *Physical Review*. He had made copies of the pages to bring back to Germany and was gratified as he flipped through the articles to see virtually no mention of nuclear research or uranium in any of the issues. He and his colleagues back home took this absence as an indication that the uranium problem was not being actively pursued by the American scientists.

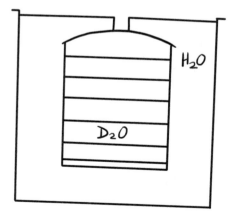

A simple diagram of the German reactor, similar to the sketch transmitted to Los Alamos by Niels Bohr. *Created by Benjamin Krapohl.*

23 | 1942

IN THE EARLY YEARS OF THE WAR, following on the heels of their successful blitzkrieg offensive in western Europe, the commanders of the German Army firmly believed that Germany was headed toward a swift victory. This belief, coupled with the assumption that America was still years away from developing its own nuclear weapon, meant that for much of the first half of the war, nuclear research, and indeed a great deal of other war-related scientific research, was kept off the Nazi military's priority list. But by the summer of 1942, the trajectory of the war had begun to change.

In June Nazi forces broke the nonaggression agreement between Germany and the Soviet government and marched into the Soviet Union. The German commanders assumed that their strategy of total, overwhelming assault would be as effective in Russia as it had been against countries to the west. They were wrong. The German tactic of "lightning war," which had quickly toppled Belgium and France, was all but impossible to repeat over the vast distances of the Russian countryside, and as Soviet premier Joseph Stalin quickly made clear, the Russians had no intention of surrendering. Through the summer and into the fall, the Nazi forces found themselves entrenched in a slow, bloody, and expensive two-front war. With resources taxed, the Nazi leaders began to look for alternative paths to victory. A search began for any project that might possibly deliver a clear and decisive advantage in the outcome of the war.

One of the men tasked with identifying these war-winning projects was Albert Speer, the minister of armaments. Speer had been peripherally aware of the nuclear research occurring under his division, and in early 1942 the program was finally brought to his full attention. The project, he was told by some of his advisers, had not been receiving the level of support it should have been, particularly given the magnitude of the possible results.

Speer arranged for a meeting to be held at the KWIP in which he brought together the nuclear scientists who had been working on the project for several years. At this meeting, Speer asked the scientists to describe the possibilities of using nuclear fission as a weapon, and to explain what the timetable might be for creating such a device. Speer later recalled that the scientists' answers were "by no means encouraging. [Heisenberg] declared, to be sure, that the scientific solution had already been found and that theoretically nothing stood in the way of building such a bomb. But the technical prerequisites for production would take years to develop, two years at the earliest, even provided that the program was given maximum support." The scientists had already largely discarded the idea of producing a nuclear weapon based on Heisenberg's overestimate of the critical mass of uranium that would be needed. When physicist Karl Wirtz had submitted a patent application for a nuclear reactor in August 1941, he had made no mention of the potential of this device in creating nuclear explosives.

Part of Speer's motivation for gaining a better understanding of the trajectory of the nuclear research program was ensuring that Hitler—who, Speer would write afterward, had a tendency to "push fantastic projects by making senseless demands"—did not get too invested in the idea of an atomic bomb if the production of such a weapon would prove impossible. While Hitler had been made aware of the possibility of nuclear weapons, his interest in the topic had so far been limited, in large part owing to Phillip Lenard, the same man who had been responsible for originating the false dichotomy of "Aryan" and "Jewish" physics. According to Speer, "Lenard had instilled the idea in Hitler that the Jews were exerting a seditious influence in their concern with nuclear physics and the relativity theory," and thus Hitler had chosen to largely ignore the topic.

Following his conversations with Heisenberg and the other nuclear scientists, Speer understood that creating a nuclear weapon in time to have any bearing on the outcome of the war was unlikely. His office formally abandoned their interest in building a nuclear weapon in the final months of 1942. Instead, the researchers were to focus on building an "energy-producing uranium motor," effectively a nuclear reactor, for power production.

While the scientists' outlook for building a war-winning nuclear weapon was not what Speer had hoped, they were able to offer a bit of good news. The enormity of the effort required to produce a nuclear weapon in Germany had only further confirmed their belief that the scientists in America were surely also several years away from accomplishing this task themselves.

The official change in focus from weapons to power production was accompanied by a change in personnel within the nuclear research group. Control of the KWIP was formally returned to the Kaiser Wilhelm Society, and, with the institute back under the control of the scientists, the unpopular Diebner was immediately ousted as KWIP director. To the surprise of no one, Heisenberg was chosen as his replacement.

After the condemnation he had suffered three years earlier at the hands of Lenard, Stark, and the other proponents of "Aryan science," Heisenberg viewed his prestigious new appointment as a resounding victory in the battle he had been waging against the pollution of physics and academic thought by political ideology. But Heisenberg had several other reasons to gladly accept the new post. As head of the KWIP, "he retained control of the ongoing atomic research and was able to protect himself and his staff from being sent to the front." The new position also came with a new academic post at the University of Berlin, and that would mean a more stable income—a relief to Heisenberg as his family was beginning to struggle under the weight of maintaining two households and feeding their growing number of children.

Despite Heisenberg's elevated position, his new authority only extended as far as the experiments at the KWIP. Diebner, now freed of Heisenberg's meddling, gathered a group of young scientists and whatever supplies of uranium, heavy water, and paraffin he could and moved his own work to Gottow, a military research installation just to the south of Berlin. Here, Diebner began work on his own series of reactor experiments. To oversee both Heisenberg's experiments at the KWIP and the work being done by Diebner at Gottow, physicist Abraham Esau was appointed the plenipotentiary of the nuclear research program.

Leading the now-divided nuclear research program, Esau struggled to maintain control over the two branches, and the growing animosity between Heisenberg—with whom Esau had his own difficulties—and Kurt Diebner was allowed to take center stage.

Building at Gottow where Kurt Diebner's research team built G-I pile, using cubes of uranium oxide and paraffin wax; and G-III, using cubes of uranium (metal) suspended in heavy water. *AIP Emilio Segrè Visual Archives, Goudsmit Collection*

Esau also struggled to ensure that both divisions of the program were given access to adequate supplies. While Degussa had continued to increase their production of uranium metal, from just 281 kilograms in 1940 to 5,602 kilograms in 1942, the supply of heavy water remained extremely limited. Although an attack by Allied forces on the Norsk Hydro plant in November 1942 had failed to affect production, it had indicated to the Germans that their interest in heavy water was no longer a secret. A second attack on the facility by a team of British-trained Norwegian commandos in February 1943 flushed hundreds of liters of heavy water into the nearby lake, and further production was effectively halted until June of that year.

Now in charge of his own research at Gottow, Diebner finally had the opportunity to try out his three-dimensional reactor design. Diebner had long believed that Heisenberg's preferred reactor geometry—layers of uranium and moderator stacked on top of (or around) one another was not the optimal arrangement for building a critical pile. He had for months been advocating for a design in which smaller units, preferably small spheres of uranium, would be distributed at regular distances throughout a moderating material. Spheres are notoriously difficult to machine, so Diebner settled on small cubes of uranium instead.

Though he was unable to acquire sufficient amounts of uranium metal, a great deal of which had just been incinerated in Heisenberg's lab in Leipzig, Diebner did have access to uranium oxide powder. Without substantial supplies of heavy water available, he settled on using paraffin wax as a moderator material. Cube-shaped wells were carved out of blocks of the wax, and these voids where then filled with uranium oxide powder, creating the "cube"-shaped elements as the pile was constructed.

The first Gottow experiment, G-I, was massive. All told, Diebner had used 25,000 kilograms of uranium oxide distributed in 6,802 cubic wells spaced two centimeters apart in 4,400 kilograms of paraffin. The experiment filled the room; the large, solid mountain of bricks of paraffin and uranium closely resembled the reactor experiment being built almost simultaneously by Enrico Fermi on the other side of the Atlantic. But while Fermi's mountainous experiment, CP-1, was about to go critical, Diebner's experiment was unsuccessful.

Back in Berlin, Heisenberg, now officially in charge of the experiments at the KWIP, continued with his preferred layer-based reactor design. Hoping to use solid rather than powdered uranium for his next experiments, Heisenberg had ordered several large, one-centimeter-thick plates of uranium metal from Degussa. Unfortunately, Heisenberg had been informed by the company that manufacturing these plates would be a far more complex process than Heisenberg had assumed and it would likely take until 1943 for Degussa to deliver them. Rather than wait for the plates, Heisenberg decided to continue moving forward with metal powder. Without access to heavy water, Heisenberg, like Diebner, turned to paraffin as his moderating material.

Heisenberg, along with von Weizsäcker and Karl Wirtz, thought that perhaps the key to producing neutron multiplication lay in both the number and thickness of the layers used. To test this theory, experiments B-III, B-IV, and B-V used seven, twelve, and nineteen layers of uranium and paraffin, respectively. None of these arrangements produced any meaningful results. Still, Heisenberg remained unwilling to concede that his layer-based design was flawed, "still mainly justifying this with the fact that it was much easier to carry out the theoretical calculations for plates" than it was for a three-dimensional lattice of spheres.

As both Heisenberg's and Diebner's experiments dragged on, Paul Harteck, who had assembled the first of Germany's experimental piles, had turned his attention away from reactor building to the problem of isotope separation. In Harteck's view, Germany had obtained more than enough raw uranium ore from Belgium to supply the reactor research. The real impediment remained the lack of heavy water. The use of natural uranium necessitated the use of an efficient moderator material like paraffin or heavy water. But if the percentage of ^{235}U could be increased, even by just a few percent, it might, Harteck correctly assumed, be possible to use a less ideal moderator, such as normal water, to build a critical reactor, skirting around the heavy water problem altogether.* Harteck took the practical view that separating out the fissile isotope of uranium and creating uranium fuel that was slightly enriched in ^{235}U was a better path forward.

With his assistant Wilhelm Groth, Harteck began working on developing methods for separating the two isotopes of uranium from one another. His first attempts involved a temperature-based method that did not show much promise. But when Groth found a description of an ultracentrifuge—a device of American origin—the pair began working with a gyroscope producer, Anschütz & Company out of Kiel, to build one. Centrifuges work on the principle that rapidly spinning—exerting a centrifugal force—on a material made up of components of different masses causes them to separate. Molecules containing a slightly heavier ^{238}U isotope will, after a lot of spinning, sink toward the bottom of the container, while the lighter isotope, ^{235}U, will float toward the top. The first centrifuge prototype was completed in April 1942, but it immediately fell apart when spun at too high a speed. But by the summer they were prepared to try their first real experiments with uranium compounds. An initial attempt, undertaken in August 1942, to enrich uranium hexafluoride

---

* Building a light water reactor only requires an enrichment of ^{235}U to about 2.5 to 3.5 percent of the total uranium content.

to 3.9 percent ^{235}U failed, but the attempt was enough to garner attention and greater financial support for the project from Esau.

In 1943 Diebner produced two more experiments. He convinced Esau to add to Heisenberg's order of mammoth uranium metal plates a smaller quantity of uranium metal cast into five-centimeter cubes. In total, Diebner received 108 of these cubes in early 1943. He also received a small amount—about two hundred liters—of heavy water. The G-II experiment was built by creating a spherical lattice of the uranium cubes suspended in place inside a mass of frozen heavy water. To protect the cubes from corrosion in the heavy water, Diebner had them each coated in a layer of polystyrene. The whole apparatus was then encased in a spherical paraffin shell. While this arrangement was awkward, and difficult to assemble, the effort was rewarded with the highest neutron multiplication of any reactor experiment yet measured in Germany.

After his success with the frozen experiment, Diebner immediately assembled another pile, with a larger and improved design. Rather than using frozen heavy water to suspend the cubes in a lattice, the G-III experiment used chains of two to four cubes strung together with metal cable. For this experiment, Diebner repurposed the 108 cubes that had been part of G-II. He also obtained some of the one-centimeter-thick metal plates that were finally being produced by Degussa for Heisenberg's layered experiments. These he had cut into five-centimeter squares, and by stacking and securing five of these squares on top of one another, he was able to add sixty more "cubes" to the arrangement. The chains of cubes were suspended from a metal lid and lowered into a cylindrical tank filled with nearly six hundred liters of heavy water.

The final Gottow experiment G-III produced a 6 percent increase in the number of neutrons measured coming out than there were going in. While still far from achieving criticality, Diebner's lattice design had unequivocally proved itself to be on the right track.

Image of the G-III reactor experiment assembly. *Niels Bohr Library and Archive, Samuel A. Goudsmit Papers (Box 26, Folder 16)*

24 | WAR IN THE SERVICE OF SCIENCE

IN THE LAST MONTHS OF 1943 the nuclear research program underwent yet another major administrative shuffle. Under pressure from Speer, Abraham Esau was removed from both of his posts: leader of the nuclear physics programs and head of the Reich Research Council's physics section. University of Munich experimental physicist Walther Gerlach was chosen to take his place.

Though before his appointment Gerlach had had very little experience in the realm of nuclear physics, he was by all accounts a "first-rate" physicist. He had a reputation as a tyrant in his own laboratory; his students and assistants feared him as much as they loved him, and he was widely thought to be "an excellent experimenter, lecturer, and leader."

Like Heisenberg, Gerlach was fiercely proud of the lofty reputation of German physics, and as the new head of Germany's nuclear program, he felt a duty to protect that reputation. Looking at the work that the Uranium Club had accomplished thus far, Gerlach was disheartened. Instead of a cohesive, well-thought-out research program, he had inherited a handful of disparate experiments being conducted by a gaggle of bickering scientists.

The scale of the program also remained minuscule. The staff at the KWIP devoted to the nuclear research program reached its highest total of fifty people in 1944, which included nineteen physicists, nine laboratory technicians, a handful of mechanics, pipe fitters, glassblowers, secretaries, and a lone engineer.

Gerlach worried that, even if Germany was ultimately victorious in the war, German scientists might emerge from their isolation to find that German science had fallen from its pedestal. He saw hope in the increasing interest of the German military in physics research, which had largely been ignited by the deteriorating certainty of a German victory. Gerlach saw an opportunity to use the wartime resources and funding to further German physics. While the scientists in Germany were more engaged than ever in the war effort, "now they used the slogan 'The War in the Service of Science,' instead of the reverse."

While the increased funding was finally providing a boost to German science, the war itself was starting to impede scientists' ability to conduct their work. A massive bombing raid on the Norsk Hydro plant by Allied planes had effectively eliminated the possibility of producing more heavy water at this facility, and a new plant had not yet been constructed in Germany. To try to salvage the last of the Norwegian heavy water that was sitting in busted tanks inside the ruins of the Vemork plant, Gerlach ordered the transfer of 614 liters of this remaining water to Germany in February 1944. The water was moved by train to Tinnsjå, where it was then transferred to the SF *Hydro*, a ferry that would take it across a deep mountain lake. Forty-five minutes into the crossing, the ferry exploded when a time-delayed device placed by Norwegian saboteurs detonated. The few precious barrels of heavy water and eighteen Germans and Norwegians were lost to the frigid winter water. The possibility of building a heavy water production facility inside Germany itself was destroyed months later by the Allied bombing of the IG Farben facility at Leuna on July 28, 1944.

Allied air raids continued to increase in frequency throughout the summer of 1944, and Berlin, where most of the nuclear research was still taking place, was becoming increasingly unsafe. On February 15, 1944, the building that housed the Kaiser Wilhelm Institute for Chemistry took a direct hit from an Allied bomb.

Determined that research should continue until the last possible moment, Gerlach made the decision in early 1944 to begin moving the sites of the different Kaiser Wilhelm Society institutes to safer locations. The exodus from Berlin began as the offices and more generalized nonnuclear equipment at the KWIP were moved, piece by piece, to the town of Hechingen, the ancestral home of the House of Hohenzollern in Germany's Black Forest. There, the scientists took over an old textile mill that they used as offices and barracks. Otto Hahn and the Kaiser Wilhelm Institute for Chemistry made a new base in the town of Tailfingen, a few miles up the road from Hechingen, while Harteck's isotope separation experiments were evacuated to the city of Celle in the far north of Germany.

Diebner's laboratory in Gottow was rapidly becoming too dangerous as well: the military installation to the south of Berlin was a prime target for Allied bombers. His experimental setup was packed up and moved to the tiny town of Stadtilm about two hundred miles south of Berlin in Germany's central Thuringia region. There, Diebner's team took over an abandoned school building and began assembling their equipment and supplies for another G-series reactor experiment.

The B-series reactor experiments that Heisenberg was still overseeing in Berlin were the last part of the nuclear program to be relocated. The underground bunker that they were working in, which had been built by Speer to serve as a successor to the Virus House, provided some shielding from the bombing above.

Spurred on by Gerlach, who was determined for the program to produce results, Heisenberg and his team constructed two more B-series experiments in 1944 in the underground laboratory. The solid uranium metal plates that Heisenberg had been waiting for had arrived and were used in the design for B-VI. The centimeter-thick layers of uranium metal were inserted into a cylindrical magnesium alloy vessel and separated from each other by spacers. The tank was then filled with heavy water, which occupied the spaces between the plates, creating the layers. While the plates were large and unwieldy, they allowed for the entire experiment to be deconstructed and put back together several times in order to test the effect of different spacings between the plates. A separation of eighteen centimeters was determined to work best when a modest increase in neutron number was detected.

Another layered experiment, B-VII, assembled in late 1944, used 1.25 tons of uranium metal plates and 1.5 tons of heavy water. This setup showed even better neutron multiplication than B-VI, though the results still fell short of those being achieved by Diebner's cube-based lattice experiments. Still, it seemed like the German scientists were slowly making progress toward criticality.

But time was running out. When Heisenberg and Weizsäcker returned from yet another conference, this time in Switzerland in December 1944, they heard about the devastating German loss in Ardennes. It was becoming clear to the scientists that the war was lost, but Gerlach hoped the battle for nuclear supremacy might still be won. Unaware of the Americans' progress in this realm, the Germans reasoned that the intellectual glory that would accompany the creation of the world's first critical pile would provide them with vital prestige as the scientific hierarchy reestablished itself after the war. German physics could still reign supreme.

Heisenberg, however, shared little of Gerlach's desperation. Even in the final days of 1944, with bombs dropping around his head in Berlin, Heisenberg insisted that panic was not required. In a letter to Elisabeth dated December 2, Heisenberg wrote: "The work at the institute is also progressing nicely right now, as is my own work. For scientific work you need the absolute precondition of a certain unpressured state of mind; with so-called industriousness, namely hasty puttering along, nothing is gained."

Recognizing that Heisenberg's group would inevitably be forced to evacuate Berlin, Gerlach initiated a search for a suitable location where the experiments might continue. On a visit to the new KWIP headquarters in Hechingen, Gerlach, who was a great lover of flowers, recalled a small village nearby that he had visited a few years prior where impressive lilacs had been blooming. The small town of Haigerloch was nestled in a deep valley of limestone cliffs surrounding the Eyach River. Surveying the small, ancient town, Gerlach noticed that next to the Swan Inn, a small cave had been carved in the rock wall at the foot of a cliff. Once the wine cellar for the cathedral perched on the bluff above, the cave now served as the inn's beer cellar.

The cellar was requisitioned by the Reich Research Council and construction on the site was soon underway to install a pit in the floor to house the reactor experiments and to build a concrete entranceway to conceal the lab's contents. The cavernous new lab space was given the tongue-in-cheek code name "Speleological Research Institute."

Back in Berlin, Heisenberg was busy assembling yet another pile. Gerlach, who could feel the clock beginning to run out, kept pushing Heisenberg to conduct a final large-scale experiment in a grand, last-ditch effort to achieve criticality. Finally acknowledging the superiority of Diebner's three-dimensional lattice design, Heisenberg had relented, and was constructing a lattice of uranium metal cubes suspended from aircraft cable into a cylindrical pit of heavy water when, on January 30, 1945, Gerlach received notice that Soviet forces were advancing fast toward the German capital. That night, he ordered Heisenberg and his team to evacuate.

Heisenberg's uranium and heavy water were loaded up into trucks, which carried the supplies first to Kummersdorf and then down to Stadtilm, where

Diebner had recently relocated his own lab. Gerlach had hoped to take the opportunity of the evacuation to pool together Diebner's and Heisenberg's combined resources and create the largest experimental pile that they possibly could. But when Heisenberg, who was already down in Hechingen, caught wind that Diebner, the second-rate army scientist who had been irking Heisenberg throughout the war, would be the one directing the construction and operation of this final large-scale experiment, he was incensed. In a letter to Elisabeth, he vented his frustrations: "In our nuclear physics group, the internal battle (Diebner vs. K.W.I) has broken out anew, probably as a result of the new wave of conscriptions and the threatening danger in the east. Maybe I will have to travel in the next days to Thuringia (Stadtilm) on account of this; I do not really like it, but perhaps it is necessary." Even in the last hours of the war, Heisenberg was unwilling to allow his rival to have access to materials that had been assigned to his group, and to possibly take the credit for building what they believed would be the world's first critical reactor, should one be achieved.

Heisenberg quickly made his way, with Weizsäcker (as always), up to Stadtilm to retrieve his material. By early February, construction had resumed on the B-VIII reactor under Heisenberg's oversight, inside the wine cellar lab in Haigerloch.

25 | BUILDING B-VIII

THE LAB IN HAIGERLOCH was little more than a literal hole in the wall, measuring just ten meters wide and reaching about three times that distance back into the darkness. The damp, moss-coated floor was set at a ten-degree angle, sloping up toward the back of the cavern, just enough of a hill that a dropped screwdriver or pencil would make a hasty dash for the front door. In the center of this room, a cylindrical pit a little over a meter wide had been dug into the floor. The walls of this pit were lined with graphite bricks that held in place a drum made of a magnesium alloy metal in which the experiment itself would be constructed.

The back of the room had been outfitted with three large metal vessels that contained the precious heavy water when it was not in use. Pipes lined the ceiling, connecting the storage containers to the experimental pit. The concrete bunker entrance that had been constructed around the mouth of the cave blocked any light from outside so that, in the cold damp of February, oil lamps lit the scientists' work as they reassembled their final attempt at a chain reaction: Major Experiment B-VIII.

Heisenberg, Weizsäcker, and Wirtz had all permanently relocated with the rest of the KWIP to Hechingen. Wirtz and Weizsäcker, who had each brought along their families, had found accommodation in small private homes, while Heisenberg, whose family was still far away in Urfeld, rented a "lovely, large room in the home of a friendly and helpful family."

Heisenberg's team and the KWIP crew found the townspeople in Haigerloch and Hechingen to be accommodating and agreeable. In such a small and isolated town, few of the people living there held much loyalty to the Nazi Party or felt any real stake in the outcome of the war; most people were simply pleased by the excitement that the arrival of the scientists had brought to their small town and were relieved by this indication that the hardships brought by the war might soon be coming to an end.

In February 1945 the Allied forces were making steady progress toward central Germany. Gerlach, who was still in Stadtilm with Diebner but in communication with the KWIP by phone, was desperate for B-VIII to produce meaningful results. For much of February and March, Heisenberg rode his bicycle each morning across the stone bridge that spanned the river to check on the progress of the reactor's construction, but much to Gerlach's frustration, he did not share his supervisor's concerns. Confident that the Americans must be suffering from the same delays that his project was, and tired of the relentless push of the large-scale experimental project, by the time Heisenberg arrived at Haigerloch "he was no longer striving for a spectacular success. . . . human problems now took precedence."

For the townspeople and scientists in Hechingen and Haigerloch, the final days of the war brought with it increasing fear and desperation. Heisenberg continued his regular correspondence with Elisabeth, but now fewer and fewer of their missives to each other were arriving at their destinations. In Haigerloch and in Urfeld, food was growing scarce. Heisenberg wrote, "Although I am getting something to eat from time to time as a gift, it remains, overall, too little. Who would have thought ten years ago that one would someday be grateful for every piece of bread someone gives you?" The hunger and uncertainty were made all the worse by the constant drone of airplanes overhead and the rumbling of bombing and artillery growing less distant by the day.

Heisenberg struggled to find ways keep the anxiety at bay. The letters he wrote home during the final weeks of the war describe distractions that he and his colleagues engaged in as they worked and waited for the end of the war to come. In February, Heisenberg and some of the townspeople arranged a small concert at a café in Haigerloch. Heisenberg played piano. He described the evening to his wife: "It may, perhaps, be crazy to undertake such a thing in these times, but at least for a few evenings one then does not talk about politics." On March 3, Heisenberg arranged a tour of the nearby Hohenzollern Castle to view the paintings by the likes of Rembrandt, van Gogh, and Leibl that filled the walls.

In the cave in Haigerloch, work slowly continued on assembling the reactor experiment. B-VIII would be, by far, the largest reactor experiment the German scientists had carried out, and it was hoped that, once assembled, its size might be enough to achieve a critical chain reaction. Following Diebner's cube-based lattice design, 664 metal cubes, each weighing about two kilograms and measuring about five centimeters on a side, were used—a total of 1.5 tons of uranium metal.

The majority of these black, heavy cubes were solid pieces of metal that had been cast or machined by Auergesellschaft and Degussa expressly for this purpose. They varied slightly in dimensions and appearance. Some, which had been roughly and quickly cast, had mottled surfaces full of craters and pockmarks. Others had been machined from rolled-out slabs, their dense faces marked by the striations left by the saw blade. A few dozen cubes had also been cobbled together by stacking five square plates that had each been cut from the large centimeter-thick uranium metal slabs that Wirtz and Heisenberg had been using in their previous B-series experiments.

To help stabilize the heavy apparatus, diagonal notches, just a few millimeters wide and deep, had been filed into the middle of two of the edges on each cube. Two pieces of aircraft cable, held in place around the cubes by these notches and crimped above and below, were used to assemble the chains. The distance between each cube in the chain had to be precisely measured. The chains were then attached to a massive annular lid that was built to cap the cylindrical, heavy water–filled pit. A small but highly radioactive radium-beryllium source could be slid into the pile from above through a hole in this lid to initiate the chain reaction.

Once assembled, the ominous chandelier of black cubes was hoisted into the air and suspended from the cave ceiling before being lowered into the pool of heavy water in the floor. The scientists held their breath, but after a few moments it was clear that the assembly had neither enough uranium nor enough heavy water to go critical. Heisenberg would later write about B-VIII in his book *Nuclear Physics*. In it he acknowledges: "The apparatus was still a little too small to sustain a fission reaction independently, but a slight increase in its size would have been sufficient to start off the process of energy production."

Heisenberg's estimations match well with Goudsmit's own back-of-the-envelope calculations made days later when the Americans arrived at the site, and with modern computer-generated simulations of B-VIII. All suggest

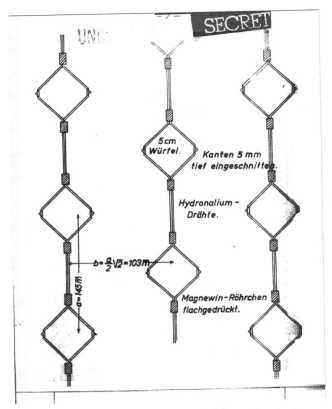

Diagram showing the assembly of the uranium cube chains. *Niels Bohr Library and Archive, Samuel A. Goudsmit Papers (Box 26, Folder 16)*

that had the pile been about 50 percent larger, criticality might have been achievable. If the experiment had been constructed earlier, or perhaps had been constructed, as Gerlach had hoped, in Diebner's lab in Stadtilm, where more material was readily available, Germany's final nuclear reactor experiment might have been a success. But in early April 1945, Heisenberg was out of time.

On April 17, Heisenberg made his usual morning trip across the Eyach River into Haigerloch. The air was still and quiet, except for the occasional low boom of distant artillery. German forces had moved through the area during the previous night as they retreated east, and in the distance, columns of smoke from burning buildings rose above the trees.

When he arrived at the lab in Haigerloch, the long strings of cubes were slowly being taken apart, one by one, and the pipes above their heads gurgled as the heavy water was pumped back into the storage vats. Together, the scientists discussed what to do with the dismantled reactor parts. Heisenberg and the others hoped to be able to continue their work on their "uranium burner" after the war was over and the dust had settled, but they worried that the Allied forces quickly advancing toward them would either confiscate or destroy the cubes and heavy water. They decided to try to hide their materials in the hopes of returning to retrieve them. To keep the locations of the materials secret, they decided that the work of hiding them would have to be done later that night, after most of the technical staff had gone home.

As they waited for nightfall, a nervous energy permeated the town as the number of aircraft flying overhead increased. Around 5:00 PM Heisenberg was back at his office in Hechingen preparing for a meeting at 7:00 PM to lay out the logistics of the evacuation when one of the technicians came running up to him.

"Can you be my witness?" the frantic man begged. "I am marrying Miss Wolfram in half an hour." The wedding was held, Heisenberg wrote in his diary, "at the town hall from 5:30 p.m. to 6:30 p.m., accompanied by low-flying planes."

In Haigerloch, late that night, with gunfire crackling in the distance, light could be seen flooding out of the small door of the laboratory bunker. Several men scurried back and forth, speaking in hushed tones as they rolled metal drums into the cave. A truck, which was backed up to the mouth of the bunker, creaked as a heavy load was deposited into the bed. The now-full drums, their contents sloshing, were rolled back out, and these were also loaded onto a truck before it headed up the road leading out of the valley.

The heavy water was hidden away in the basement of an abandoned mill a few miles up the road. The uranium cubes were driven up one of the surrounding hills to a farmer's field that had recently been plowed. A pit less than a meter deep was dug into the ground, and a wooden crate was lowered down. One by one, the cubes were placed into this makeshift subterranean vault. With the lid secured, the hole filled in, and their treasure now buried, Heisenberg and his team walked away from their uranium.

Seventy years later to the day, Tim Koeth stood in the same field, Geiger counter in hand, comparing his location, based on the arrangement of rooftops in the valley behind him, to old photographs taken of the Alsos Mission men as they painstakingly removed the cubes from the ground. As he positioned himself on what he approximated to be the correct patch of earth, he squatted and scanned the dirt with his radiation meter, hoping to hear the familiar clicking of radioactive metal. But the meter remained silent.

As Tim was scanning the newly plowed field, an older gentleman approached. The man, who was apparently the farmer on whose property Tim was currently standing, didn't speak any English, but Tim tried his best through gesticulations and the photographs in his hand to explain his presence. But when the man caught sight of Tim's yellow Geiger counter, no further explanation was required. Tim was not the first to come to this field hoping to find buried treasure left behind by the Alsos Mission. The man chuckled to himself, shook his head, and turned to walk away, leaving Tim to his hunt.

26 | FARM HALL

MOST OF THE SCIENTISTS and technicians working in Haigerloch, Hechingen, and nearby Tailfingen planned to remain where they were and surrender to the Allies upon their arrival. Heisenberg, however, was determined to get to Urfeld and be with his family as the war came to an end. On April 19, 1945, at 3:30 AM, in the freezing cold dark of the early spring morning, Heisenberg got on his bicycle and set out on his 270-mile journey home.

The chaos of this late stage in the war made the road extremely treacherous. Elisabeth describes her husband's journey:

> Everything was beginning to disintegrate; he encountered bands of marauding, tattered figures speaking foreign languages, who had been released or had escaped from some prison camp or from forced labor, and who were now roaming through the countryside plundering; he saw groups of 14- and 15-year-old boys who had been drafted and abducted [at] the last minute, and who were now camping along the side of the road, crying, hungry, and lost, not knowing what to do; he met hordes of soldiers of the most varying nationalities, going somewhere, some to the east, others to the west or north, without a plan, exhausted, and threatening.

Heisenberg did his best to avoid interaction with anyone and spent the daylight hours of his journey hiding in ditches along the road. On the second day of his journey, he was unable to avoid detection by an SS officer when forced to pass through a checkpoint. The officer, noticing the nervous and tired man, who was not wearing a military uniform, waved him over and asked to see his travel papers. Heisenberg nervously handed over the travel permit that he had issued to himself prior to setting out. The stakes at this moment, Heisenberg knew, were particularly high. If the papers didn't pass scrutiny, Heisenberg ran the risk of being labeled a deserter. He was familiar with the many rumors circulating along the retreating front of men suspected of having abandoned their units being summarily sentenced to death and shot in the road.

The inspecting officer looked at Heisenberg's papers, and, puzzled by the odd signatures, moved to take them to his superior in a nearby tent. Heisenberg's heart sank. Thinking fast, he reached into his pocket, where he had stashed a pack of American cigarettes that had been given to him several nights before by a friend in Hechingen as a parting gift. Cigarettes, particularly American cigarettes, had become a rare and precious commodity.

He asked the soldier if he smoked. The young agent nodded, and Heisenberg pressed the packet of Pall Malls into his hand. The guard took the bribe and moved aside, allowing Heisenberg to pass.

On the morning of April 21, Heisenberg reached the town of Kochel south of Munich. His house and family were just on the other side of the Kesselberg mountain. At 9:00 AM he began his ascent of the massive hill, and by 11:30 AM he had reached the summit and could see the lake and town below.

The house in Urfeld was little more than a two-story cabin. A sloping roof and a balcony jutted out from the center of the top floor and a small wooden gate opened onto a large patio surrounded by a stone wall. Looking out the front window that afternoon, Elisabeth was stunned to see her husband climbing the road to the front door: "And finally, suddenly and unexpectedly, I saw him coming up the mountain, dirty, dead tired, and happy."

For a few days, the Heisenberg family settled into a calmer routine. When news of Hitler's suicide reached Urfeld, Elisabeth wrote that they "fetched the last bottle of wine—we had actually been saving it for the baptism of our daughter—from the cellar, and drank it with tears of relief and deliverance."

But the war was not yet over for the Heisenbergs. Days later, an American colonel, wearing glasses and with a Greek letter alpha and lightning bolt emblazoned on his uniform, appeared at their front door. It was a gorgeous, clear blue morning in the snow-covered lake town, as Germany's top physicist was escorted to a jeep that would take him back to Alsos headquarters in Heidelberg for questioning.

In Heidelberg the Alsos team had cleared out of one of their villas on the hill near the Philosophers' Walk to make room for their new prisoners. When Heisenberg arrived, he joined several of his colleagues from Hechingen and Tailfingen who had been arrested by the Alsos team the previous week. A few days later, Gerlach and Diebner, who had been picked up by Carl Fiebig and Gerry Beatson in Munich, also joined the ranks of the scientist-prisoners.

All the captured scientists were kept separately from one another but were given a comfortable room in which to live while they awaited the next phase of their imprisonment. A soldier was posted outside each man's door, and as it turned out, the only unit available for guard duty was a Black regiment. The German scientists who "had perhaps lived too long under the myths of Aryanism" were more vocal in their displeasure over the identity of their guards than by their confinement. Needless to say, the Alsos men were in no hurry to make any changes.

By this point in the war, Goudsmit had grown accustomed to interrogating scientists whom he had once seen as his colleagues, mentors, and even friends, but the interrogation of Heisenberg felt different—more personal. After all, Goudsmit had hosted the German physicist at his own house only weeks before the war had begun, and he had begged Heisenberg to stay in the United States rather than return to work in Germany. Heisenberg's stubborn refusal had led the pair directly to this moment: the German prisoner and the American interrogator.

Goudsmit had been planning his questioning of Heisenberg for months, playing out different scenarios in his head for how the German might react. By the time he walked into the room where his old friend was being held, Goudsmit knew exactly what he would say. Heisenberg, on the other hand, was taken completely by surprise when he suddenly recognized the man dressed in the uniform of a senior US Army officer.

"My dear Goudsmit!" Heisenberg exclaimed, jumping up and holding out his hand. The American did not offer his hand in return.

Their conversation lasted several hours and spanned many different topics—Heisenberg seemed utterly unaware that over the past few years, the scales of the physics world had tipped decidedly in America's favor, and that the Germans had been left in the dust.

"If American colleagues wish to learn about the uranium problem," Heisenberg offered, "I shall be glad to show them the results of our researches if they come to my laboratory."

———————

By the end of May, the interrogations of the German scientists had largely been completed by the Alsos team, but the fate of the captured men remained undetermined. While it was clear that the Germans had not gotten anywhere close to a nuclear weapon, Groves worried that during the course of their interrogations, some of them might have begun to piece together that the American program was much further along than their own. If the Germans were to be released, the Americans ran the risk of these men spreading their suspicions to the wider science community. Or worse, the Soviets, who had made no secret of their own nuclear ambitions, might scoop up the German scientists themselves and force them to work on the Soviet nuclear program. Some German physicists, Groves knew, had already been spirited away behind Soviet lines.

While releasing the German captives was not an option, determining where to keep them quickly became problematic. After a few weeks of waiting in Heidelberg, they were transferred to Versailles, where they were placed in the "Dust Bin," the name given to the internment center for important civilians who had worked for the Nazi regime. Conditions were terrible. Unlike in their comfortable accommodations in the mansion in Heidelberg, the men and women being held at this center were treated as real prisoners of war. So great was the scientists' distress over their new situation that after just a few days, Pash and his team moved their captives yet again, this time to a house in Belgium, until they could determine what to do with Heisenberg and his cohort on a more permanent basis. Eventually, Michael Perrin and the British government stepped in with a solution: an estate about forty miles outside London that had regularly been used as a "safe house" by British Foreign intelligence, known as Farm Hall.

In all, ten German scientists were transferred to England on July 3, 1945. Guest Number 1, Heisenberg, was accompanied by his two right-hand men, Carl von Weizsäcker and Karl Wirtz, along with the other two leaders of the nuclear program, Walther Gerlach and Kurt Diebner. Two of the younger scientists, Erich Bagge and Horst Korsching, who had been working under Heisenberg at the KWIP, had also been selected for interrogations and were imprisoned with their mentor. The remaining three guests had not been directly involved in the final phases of the nuclear program but had all been deemed potentially valuable assets. Paul Harteck's work on isotope separation had earned him a spot, as had Otto Hahn's early work on fission. Finally, while he had had no direct contact with nuclear work itself, Max von Laue, the 1914 Physics Nobel laureate, had been working with Hahn in Tailfingen at the time of their capture and was forced to join his colleague in England.

While the prisoners were not permitted to leave the estate, they were captives in a "golden cage." They were given free run of the grounds and tennis court, and Heisenberg was even provided with a piano. For entertainment, the scientists spent the summer taking turns giving lectures on various topics in physics. What they did not know was that before their arrival, the house and grounds had been bugged, and full records were being kept of their conversations. In one particularly ironic exchange, Diebner was recorded wondering aloud if their captors might be listening to them, to which an amused Heisenberg, once again underestimating his American and British counterparts, laughingly replied, "Microphones installed? Oh no, they're not as cute as all that. I don't think they know the real Gestapo methods; they're a bit old fashioned in that respect."

The conversations recorded between the scientists during that spring and summer included complaints about lack of contact with their families back home, half-baked plots to jump over the garden wall and try to find old friends in nearby Cambridge, and the writing of a song entitled "The Farm Hall Nobel Prize Song," set to the tune of "Studio auf einer Reis," whose first two stanzas went:

> Detained since more than half a year.
> Are Hahn and we in Farm Hall here.
> If you ask who bears the blame,
> Otto Hahn's the culprit's name.

The real reason, by the by,
Is he worked on nuclei.
If you ask who bears the blame,
Otto Hahn's the culprit's name.

But perhaps the most interesting exchange between the scientists was recorded on August 6, 1945, when, shortly before dinner, the scientists were informed of the United States' nuclear attack on Japan. Otto Hahn was the first to be told by one of the guards. He was immediately extremely distraught—he felt that, as the discoverer of fission, he was partially responsible for the terrible consequences his discovery brought. But "with the help of considerable alcoholic stimulant he was calmed down and [he] went down to dinner where he announced the news to the assembled guests."

The first reaction from the scientists was stunned silence. Heisenberg was one of the first to speak, launching into an explanation of how the news must have gotten something wrong—it simply could not have been an actual atomic weapon: "All I can suggest," he exclaimed, "is that some dilettante in America who knows very little about it has bluffed them in saying: 'If you drop this it has the equivalent of 20,000 tons of high explosive' and in reality doesn't work at all."

But slowly, the scientists began to put the few pieces of information offered by the BBC announcement into place, and the possibility that the atomic bomb was indeed real began to coalesce.

"If the Americans have a uranium bomb," said Hahn to his colleagues, "then you're all second raters. Poor old Heisenberg."

———————

As the dust settled in the days and weeks after the bombing of Japan and the end of World War II, the players in Germany's nuclear research strove to construct a new narrative around their involvement in German wartime science. Many of them, none more fervently than Heisenberg, "made failure into a virtue" and would later claim that they had never actually intended to create an atomic weapon—that their pursuit of nuclear power had been solely aimed at generating energy, not an explosive. In his book on the topic of nuclear fission, written in the years after the war, Heisenberg takes care to mention the uses of atomic energy in "power stations, stations for generating and transmitting heat, and naval engines, driven by nuclear energy," while conspicuously leaving out any mention of the uses of fission in warfare.

Whether the German Uranium Club intended to build a nuclear weapon remains a matter of historical debate, but what is clear is that the course that the German scientists and Nazi government and military took in developing and managing their nuclear program would not have resulted in a nuclear weapon during the course of the war.

But for many of the scientists who had been forced or had chosen to leave Germany as the Nazis gained control, the success or failure of their German colleagues' research was entirely beside the point. These men had chosen to stay, to actively participate in research for the Nazi war effort, and in doing so had made themselves complicit in the violence, cruelty, and havoc wrought upon Europe by the Nazi regime. In a letter to her friend Otto Hahn in June 1945, during the time that he was interned at Farm Hall, Lise Meitner shared her view of the German nuclear scientists' role in the war:

> You all have worked for Nazi Germany as well and never even tried to put up a passive resistance either. Certainly, to buy off your consciences you have helped a person in distress here and there, but have allowed millions of innocent people to be slaughtered without making the least protest. I must write this to you, because for your sake and for Germany's so much depends on your understanding what you have allowed to happen.

Following his interrogations of the German scientists at Heidelberg, Samuel Goudsmit had made his way to Berlin. The building that had housed the KWIP—a white, dense, neoclassical structure—had survived the war unscathed. An American soldier posted at the building's entrance told Goudsmit that there wasn't much left to see inside the building—all the wiring and pipes had been stripped out by the Russians. The few odds and ends that had been left behind had been thrown into the yard behind the building. Inspecting this "junk," Goudsmit found traces of the work that had taken place there over the past several years—among the bits and pieces of broken instruments he found a handful of pressed uranium oxide blocks.

Goudsmit walked down to the subbasement where the Germans had built their experimental pit to hold their reactor that never achieved criticality. As he stood alone in the dark and damp, Goudsmit would later write, "In the dim light [I] thanked God for the great privilege of being permitted to see with my own eyes and in a language I could understand, the physicist's symbol of the defeat of Nazism."

27 | THE 400

BY MID-MAY 1945, Alsos had captured all of the top German nuclear scientists along with the supplies of heavy water and uranium metal cubes that had been uncovered near Haigerloch. What they had not yet been able to secure were the additional reactor materials that had been used by Kurt Diebner at his relocated lab in Stadtilm before both he and the materials were whisked away by SS agents.

In early April, just days before Alsos arrived at the site, the basement lab in central Germany, which had been largely quiet since Heisenberg had taken his own stock of supplies down south, was suddenly abuzz with activity. Agents from the intelligence wing of the SS arrived one morning and hastily began loading the lab's supplies of heavy water, paraffin, pressed uranium oxide briquettes, and a small amount of radium into three large trucks, along with a sizable supply of uranium metal in the form of five-centimeter cubes.

Along with his reactor supplies, Kurt Diebner himself was placed into protective custody by the agents. The plan was to move both the scientist and his materials east, to Weimar, and then possibly on to Innsbruck, Austria, where the dwindling Nazi forces had been rumored to be gathering and preparing to mount a last stand against the ever-advancing Allied forces. Presumably, they hoped that by bringing Diebner and his uranium behind the lines with them, the scientist might be able to conjure up a last-minute scientific miracle to aid the Nazi forces.

But there was to be no miracle, nor any grand last stand. The loaded trucks left Stadtilm and slowly made their way south, but within days of their departure, Soviet forces encircled Berlin and shortly thereafter, news of Hitler's suicide began to spread.

On April 25, the same day Alsos was digging up uranium cubes in Haigerloch, the SS convoy located Walther Gerlach in a small town just south of Munich. With thoughts of reconstructing any kind of experiment long gone, Gerlach and Diebner were now motivated by the same thought that had driven Heisenberg to have his own uranium hidden below ground; the war was surely lost, but with stores of uranium and moderator hidden away, the scientists might be able to resume their work in a few months.

The materials from Stadtilm were split up, and two trucks were sent north. Gerlach returned to his home in Munich, while Diebner rode in one of the northbound trucks to Schöngeising, where Alsos members Carl Fiebig and Gerry Beatson would find him a few days later. The scientist and supplies were then transferred back to the Alsos base at Heidelberg.

A third truck that carried the majority of Diebner's stash of uranium cubes was not sent north but south, to the town of Garmisch, which by coincidence was only a few miles from Urfeld, where Heisenberg had just reunited with his family.

On April 26, Karl Bischoff, a twelve-year-old boy, was bicycling home when he saw an army truck zip into an alleyway between a house and a woodshed. As he watched, the two soldiers driving the truck got out and hurried away from the vehicle, leaving it unguarded. Karl, who had grown accustomed to the patterns of military operations passing through town over the past several months, thought this sight was particularly odd. He quickly alerted his three friends, Michael and Ludwig Hosp and Weiß Ferdl, and the boys took turns keeping an eye on the seemingly abandoned vehicle. After three days, when no one had returned to the truck, the curious boys could not contain themselves any longer. One stood guard at the entrance of the alleyway while the other three lifted the tarp covering the back of the mammoth vehicle and peered inside. The truck was loaded with dozens of small but very heavy wooden boxes. A hammer was produced, and the boys pried one of the crates open. To their disappointment, instead of food, or even some sort of mysterious and potentially explosive military implement, inside the crate they found a handful of black hunks of metal.

Disappointed in the result of their stakeout, the boys carried one of the cubes with them as they hopped back on their bikes and headed down to the nearby Loisach River in search of another fun diversion. They were playing with rocks on the bank, tossing them at the water as kids the world over are apt to do, when suddenly one of them picked up the heavy cube and lobbed it at the concrete bank barrier. To everyone's surprise and utter delight, when the cube ricocheted off the wall, bright white sparks were sent flying into the air.

The boys spent an exciting afternoon throwing the cube and watching these sparks scatter. Eventually, though, this new game grew dull, and after pocketing a couple of the cubes for their bedside tables, the boys headed home and quickly forgot about the heavy metal objects in the abandoned truck.

But the children were not the only ones who had taken notice of the truck or its unusual contents. Anton Biersack, a local hockey legend in the Bavarian town, also took note of the strange rocks that the local boys had been playing with. The cubes, he reasoned, must be valuable since they were being transported in a military truck; and the sparks that flew from them so easily clearly pointed to some sort of special properties. He and some friends quietly stashed the crates away until they could decide what to do.

In mid-August when news of the new and terrible weapons that had been dropped on Japan spread throughout the world, and the name of the material that powered it was on everyone's lips, Biersack and his friends realized what they had taken: uranium. A scheme was hatched, and the beginnings of a black-market trade began to take shape.

When German territory had been divided among the Allies, creating so-called zones of occupation, the United States had been given control over most of the south, sharing a small sliver on the west with France, while Great Britain gained control over the northwest and the USSR was given power over the northeast. The city of Berlin was similarly divided among the four Allies. After VE day Allied Control Commissions were formed to serve as governing bodies for the German territory. A battery of new rules and restrictions for German citizens were quickly instituted, designed to quell any possible rebellion from remaining Nazi loyalists. Among these restrictions, the new legislation removed from public office any individual who had participated in the Nazi regime, forbade the display or wearing of Nazi uniforms and symbols, and prohibited German citizens from possessing any war-related materials.

When the Soviet representative of the Allied Control Commissions suddenly withdrew from the organization in 1948, the governance of Germany was split in two: East Germany was overseen by the Soviet Control Commission, while the West was overseen by the Allied High Commission, also called the High Commission for Occupied Germany (HICOG).

Beginning around 1948 HICOG intelligence officials began to catch wind of a small but growing trade in radioactive goods, as "jobbers" passed around stocks of materials scavenged from the remains of the German nuclear program. The first confirmation that this rumored market was real came in late 1948 when a uranium cube, five centimeters on a side and weighing about 2.5 kilograms, "was obtained by the U.S. Military Police in Frankfurt from a man who bought it on the black market in Munich." The origin of the cube was traced back to Garmisch, and Anton Biersack, along with eight of his accomplices, were arrested on April 22, 1948, and charged with illegal possession and sale of war materials.

A trial ensued, and the cube was confiscated and transferred to the headquarters of the Twenty-Seventh Military Police's Crime Laboratory. Biersack and his friends were found not guilty of the charges because, as the presiding judge pointed out, a block of uranium was still a long way from a nuclear weapon, but their confiscated cube was not returned. Instead, in October 1954, the French, English, and American representatives of the Scientific Research Division of the Military Security Board approved a request made by Professor A. Neuhaus, the chair of the Institute of Mineralogy and Petrology at the University of Bonn, to transfer the cube as a permanent addition to their collection. It is still on display in the university's mineral museum today.

While Biersack and his buddies had been arrested for selling one cube, dozens, and perhaps hundreds, more had made their way onto the black market. And it was not just cubes up for sale. Other radioactive materials—from uranium metal powder to uranium oxide to brightly colored uranium salts, all scavenged from the abandoned labs and warehouses that the German physicists had left behind—were being sold.

Every few weeks, a letter or two would find its way to officials at HICOG, sent by would-be uranium salesmen. These letters read with a kind of desperate earnestness, and in the text, the authors often revealed an obvious and complete lack of understanding of what exactly it was they were trying to sell. Many of these letters claimed that these blocks were not just uranium but specifically

^{235}U, a name that they had no doubt heard on a BBC news bulletin. But, as the Americans were now aware, all uranium available in Europe was natural abundance: mostly ^{238}U. As a memo sent from the HICOG Intelligence Division pointed out, the "material cannot contain more 235 than normal unless stolen in USA."

Separating fact from fiction in these communications was not always so straightforward. In July 1949 one of these letters appeared, claiming that a cube of metal weighing approximately two kilograms, measuring five centimeters on a side, and described as "black with a green sheen" was available for purchase outside Rome. In the letter, the original provenance given for the cube was the made-up name of a nonexistent town but which seemed to contain hints of the truth, like the garbled result of a game of telephone: "Product of German Atomic Pile of Heigenlche Nel Wurttemburg." The writer also helpfully suggested that if the identity of the cube was in question, the US authorities might consider asking Frédéric Joliot-Curie in Paris for assistance.

Some of these letters had a darker tone than others, and the offer to sell uranium would sometimes be accompanied by a thinly veiled threat that if the United States was not willing to purchase the uranium, it would be sold to the Soviets instead. A letter to David Lilienthal, the chair of the US Atomic Energy Commission, detailed four blocks of uranium. "At least one block of this uranium," the letter warns, "has been sold, and in all probability is in the hands of people who are not considered over-friendly to the U.S." While the amount of uranium in question was, by this time, too small to have any impact on either the US or Soviet nuclear programs, the added Cold War drama only served to fuel anxieties as these notices continued to appear.

As time went on, the peddlers on the uranium market grew bolder. In August 1950 a letter offering the sale of a small amount of uranium oxide powder was sent directly to the scientists at Los Alamos. In April of the next year, a letter signed by "a friend of true democracy" arrived at the White House addressed to President Truman.

The work of sorting through these letters fell primarily to two men. Gordon Arneson was serving as special assistant for atomic energy for the secretary of state when the bulk of the black-market uranium trade was occurring, and many of these missives made their way to his office. Clarence Wendel, a special consultant for the Military Security Board, was on the ground in Germany, and together with Arneson and the Atomic Energy Commission (AEC), handled each of these cases and provided the United States' official response.

While the US nuclear program was still hungrily searching for new sources of uranium, the amount of money demanded for each cube was always far above what the AEC was willing to pay—often completely fantastic sums. A communication intercepted by American intelligence from a Hans Palt in Vienna notes that the going price for a 2.29-kilogram uranium cube on the black market had gone up from the $15,000 requested by Biersack to 4.5 million Swiss francs (just over $1 million).

In an attempt to rein in some of the growing consternation surrounding these communications, in April 1949 an official memo was distributed to Wendel, Arneson, and the State Department laying out the AEC's position:

a. The Commission is aware that small quantities of ex-German uranium metal exist in the black market area of Europe.

b. The portion of this German uranium metal still available in Europe would make no essential contribution to any atomic energy program in any country.

c. The Commission feels that in most cases samples should be obtained, if available, in order to confirm the fact that the particular metal available is German metal and not U.S. or U.K. metal which would represent a security violation of some importance.

d. The normal "market price" for uranium metal is around $6.00 a pound.

While the amounts of uranium circulating in Europe were relatively small, it remained difficult to know just how much there might be. Most of the reports remained unverified, and it was impossible to tell if each communication was discussing a separate stash of cubes or if many people were trying to sell the same few. By early 1951 many of the reports began to tell a similar story: "It appears that the members of the ghostly gang are gradually taking shape, and that the present location of their source of supply is in Switzerland." It was estimated that this supply contained about a thousand kilograms of uranium metal in the form of 2.5-kilogram cubes—about four hundred cubes in total.

It was not until 1952, however, that another one of these cubes surfaced. On December 12, 1952, two Berliners, Helmut Goltzer, a sales agent, and his girlfriend, Gisela Nitzke, were arrested for possession of a 2.5-kilogram block of uranium metal. Their arrest and the confiscation of their cube was reported in the news: "According to the accused, they got it from a 'foreigner' and it came from the uranium mine at Aue, near Leipzig, in the Russian Zone. German police, however, believe it came from the war time atomic research station of Haigerloch." Apparently, the couple had been trying to sell the block for some time with no success and had finally tried to sell it to the Max Planck Institute for Physics when they were caught.

––––––––––

After the downfall of the Nazi regime at the end of the war, the remaining scientists at the Kaiser Wilhelm Society and its individual institutions were faced with deciding how to proceed. Over the previous twenty years, the influence of National Socialism on the society had been unmistakable, and many of the ethical lines that had been crossed by scientists in pursuit of Nazi ideals were irrevocable. So far had the society sunk during the Nazi regime that its final president, Albert Vögler, had, like so many Nazi true believers, committed suicide just before the Allies arrived.

The war itself had also taken a toll. Many of the society's scientists and technicians had been killed, not to mention the large number of Jews and others who had been forced out of their posts in the 1930s. The buildings of the society, many of which had been located in heavily bombed Berlin, were also damaged or destroyed.

In 1948 the remaining scientists at the Kaiser Wilhelm Society, in a bid to reestablish Germany's scientific credentials, formed a new society, under the leadership of the eighty-seven-year-old Max Planck. Otto Hahn, since released along with his fellow captives from Farm Hall, became the Max Planck Institute's first president. Slowly, German science was rebuilt and the Max Planck Institute remains a beacon of fundamental science research today.

But in the late 1940s and early 1950s, despite the rebranding and fresh start, there was little doubt that the physicists working in Germany had found their place in the postwar research landscape drastically changed. Now, with the new restrictions imposed by the Allied Control Commission, and later HICOG, and their supplies of uranium and heavy water largely confiscated or scavenged, the nuclear physicists found themselves without any nuclear material to study.

The public arrest of Goltzer and Nitzke provided a sliver of hope. After the trial, the cube in question had been transferred to the HICOG Operating Facilities Division for disposal, but in February 1954, two letters were received from scientists at the new Max Planck Institute for Physics, seeking to obtain the cube for research.

The first letter was sent by a Dr. Forstmann who began his appeal in oddly imperious fashion: "In the name of Professor Dr. Otto Hahn." Forstmann suggested that the uranium cube in question should be returned on the grounds that it had been "stolen in 1945 from the Kaiser Wilhelm Institut fuer Physik (Max Plank Institut) which had been moved to Haigerloch." He goes on to complain that the lack of material had impacted important work: "Nobel prize winners Prof. Dr. Werner Heisenberg, Göttingen, and Prof. Dr. v. Laue, Berlin, both internationally famed scientists would appreciate the uranium cube being turned over to them for their scientific research work, because they have been unable to obtain another one so far." The notion that the cubes were stolen scientific objects rather than war material confiscated by the Allied forces echoes the sentiments, apparently unchanged by their internment, of Heisenberg and several of his colleagues that their work was somehow distinct from, and that they themselves were not directly complicit in, the Nazis' warfare agenda.

A similar, though somewhat more evenhanded letter was sent by Max von Laue himself on February 21. While less bombastic, this letter also suggests that the cube in question had been "stolen."

In the end, unlike the appeal from the University of Bonn, the scientists' requests were not granted. A letter addressed to Clarence Wendel explained the reasoning behind the denial: "This matter has been discussed informally with the British and French representatives on the Committee, and they are in agreement with me that it is unwise for us to donate the uranium in question to a local research institute, and to license the research activities which would probably ensue. The basis of this decision is entirely political." The Allied forces worried that allowing research to resume in Germany at the same institution that had carried it out during the war would be perceived badly on the larger global stage. Heisenberg would not be getting any of his uranium back.

Beyond these two cubes, little of the German uranium ever resurfaced, though the leads received by the AEC grew increasingly fanciful. In one instance a "uranium parcel," which had been placed in a bank vault in Zurich and was retrieved through use of a provided secret code by American agents, turned out to be only a few grams of uranium salts wrapped up in paper. Most other leads that were followed led to similarly underwhelming discoveries.

It remains largely unclear exactly how many black-market cubes existed in central Europe in the years following the war. If there was indeed a large stash of hundreds of cubes, what might have become of them is equally unclear. A 1953 letter from Gordon Arneson to the US embassy in London provides a possible answer:

> We read with interest the enclosure to your letter of July 27 concerning the Russian purchase of "several parcels of uranium worth $6 million." The thought immediately comes to mind that the material is similar to that referred to in the Department's circular airgram of February 25 1953, 4:25 p.m., on the black market in uranium cubes in Western Europe. It was noted in this aerogramme that usually at the time an offer is made to US of a kilogram or two of U^{235} for a million dollars or so, a threat is delivered that the material will be sold to the USSR unless the US purchases it. It seems that at last such a threat has materialized.

It would seem that Diebner's cubes most likely made their way into the Soviet Union, to be used as fuel in their rapidly expanding nuclear program.

28 | PAPERWEIGHTS

MOST, IF NOT ALL, of the uranium metal cubes that had once formed the basis of Germany's nuclear research program, and that still survive today, have found their way to their current locations after being picked up at some point in their journey as a souvenir—a scientific spoil of war. All the cubes from the Haigerloch site had been whisked away by military transport, but when a small cache of cubes from Diebner's stash found their way to the Alsos base at Heidelberg, the mission's scientists and soldiers saw an opportunity to purloin one of these objects for themselves. For some, this opportunity was irresistible, and the souvenir-scrounging Alsos members picked their favorite radioactive bricks off the pile. Samuel Goudsmit, ever the collector, took two.

Goudsmit was pleased with his trophies, and these cubes found their way onto his desk in Paris. By themselves, the lumps of metal are novel but largely useless for any purpose beyond holding down paper. This is just what Goudsmit did. Visitors to his office in the summer months of 1945 found two black bricks perched atop piles of reports and documents.

One of these visitors was Paul Harteck. Though he was a member of Germany's nuclear research program, Harteck had spent the majority of the waning years of the war working on the development of a centrifuge in hopes of achieving isotope separation, leaving Heisenberg and Diebner to squabble between themselves over the uranium blocks. So when Major Russell Fisher finally captured Harteck in Hamburg in mid-May 1945 and brought him to Goudsmit's office for questioning, he did not immediately recognize the dark objects sitting on the American's desk.

Harteck was one of the few German scientists whom Goudsmit had not met before the war. Unaware that Alsos had already been to his laboratory in Celle—and therefore knew most of the details of what he had been working on—as he sat down in Goudsmit's office, Harteck was at first haughty and evasive. Goudsmit tried to coax him into talking by asking a few questions that were only tangentially related to the uranium research. He knew that Harteck's group had been working on generating complex molecules centered around a uranium atom.

Harteck admitted that his lab had been trying to create uranium compounds, though he did not go so far as to confess that this work was being done in the hopes that they might find a compound that would enable the separation of uranium isotopes via gas diffusion. Their work, Harteck elaborated, had not been very successful.

Harteck glanced around Goudsmit's desk looking for knickknacks or trinkets that could be used as proxies for atoms to illustrate the large and complex molecules that he and his colleagues had been working to synthesize.

Seeing a dark-gray paperweight perched atop a stack of reports, Harteck reached to grab the small two-inch cube, saying, "Now let us assume that that represents uranium." As he tried to lift the paperweight and move it into position, he was caught by surprise by its unexpected weight.

Goudsmit relished seeing the color drain from Harteck's face as the pieces began to fall into place in the German scientist's mind, and any traces of superiority evaporated from his eyes. He shouted, "But this *is* uranium!"

Harteck suddenly recognized what the cube was and where it had come from. If the Americans had gotten that far, he quickly surmised, they must already know almost everything. Stunned, Harteck was suddenly much more willing to share everything he knew as Goudsmit returned to his original line of questioning.

The presence of the uranium cubes on Goudsmit's desk was, sadly, relatively short-lived. After a visiting officer let slip back in Washington that he had seen these objects in Goudsmit's office, Groves, ever vigilant about security, ordered the Alsos team members to put them out of sight. "It was too bad," Goudsmit bemoaned. "They made such nice paperweights."

Years later, visitors to Samuel Goudsmit's office at Brookhaven National Labs, in Upstate New York, would frequently see a black block perched on top of a stack of papers on the desk.

Samuel Goudsmit at his desk at Brookhaven National Labs. *Courtesy of Brookhaven National Labs*

While it is likely that the lion's share of Diebner's stash of uranium cubes made its way into the uranium-hungry USSR, Pash and his team had packed up and shipped out 659 of the cubes from Haigerloch to Paris. These cubes appear in the historical record only one more time after their capture, in a memo written by Goudsmit in May 1945. The missive, entitled "Activity of the Material," reports to Major Furman the findings of Goudsmit's assessment of the radioactivity emanating from the inch-thick wooden boxes in a Paris depot that contained the captured cubes. Goudsmit noted that while his equipment was limited for making these measurements, particularly in detecting the presence of any dangerous fission products that could be present, "the type of experiments done makes any appreciable additional activity very unlikely." The cubes were relatively innocuous, he concluded, and could be safely transferred.

As for where the cubes were headed next, Groves provides a hint in his memoir:

> About one and a half tons of small metallic uranium cubes were dug up from a plowed field just outside town. These, too, were quickly dispatched to Paris. Both water and uranium were then shipped to the U.S., to be disposed of by the Combined Development Trust.

Sometime that summer the boxes at the Paris depot were loaded onto a ship and made their way back to the United States.

PART III

GIFT OF
NINNINGER

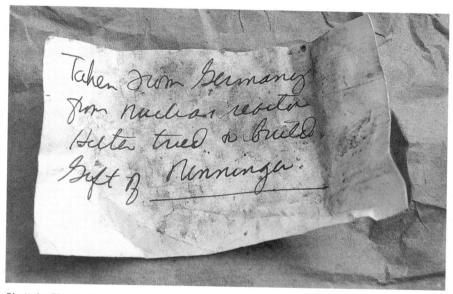

29 | FINDING NINNINGER

SITTING AT HOME IN FRONT of his computer after his meeting in the university parking lot with Mary Dorman, Tim Koeth worked to make sense of the note that had accompanied the cube he had just been handed. "Taken from Germany" clearly referred to the Alsos Mission and the removal of nuclear material from Germany. "The reactor Hitler tried to build" indicated Heisenberg and his nuclear program. It was the last clause that was giving Tim trouble: "Gift of Ninninger." Tim tried searching the Internet for the name, but nothing helpful came up.

Just a few weeks later, Tim found himself in southern Virginia, at an annual gathering of amateur physicists. This event, which takes place outside Richmond in the front yard of the eccentric man who has been organizing the get-together for years, is focused on the building of basement and backyard fusion devices. The gathering draws all sorts of characters, spanning the full breadth of the mad-scientist spectrum, who travel to Virginia to show off the homegrown experiments that they have been working on over the previous year. No stranger to basement science, Tim is a frequent attendee.

Before the real science begins each year, these weekends always kick off with a unique swap meet. Plastic foldout tables are set around the yard and festooned with dazzling arrays of cables, power sources, and electronic trinkets of all kinds, electrical and radiological novelties, and (of course) books.

As Tim was picking through a milk crate of old musty volumes, one of the books suddenly caught his eye. A man wearing a hard hat and holding a yellow rock in a desert landscape adorned the faded yellow book jacket under the title *Minerals for Atomic Energy*, by Robert D. Nininger. Three *n*'s. The note that came with his cube spelled the name with four, but this was a coincidence too unlikely to ignore. Tim was certain that the author of the book he was holding had to be the same man he had been futilely scouring for on the Internet. Tim bought the book for ten dollars.

Now that he was armed with a first name, and correct spelling, the story of Nininger and Tim's uranium cube quickly began to flesh out. Robert D. Nininger, it turned out, had been a resident of Rockville, Maryland, just a twenty-minute drive around the DC Beltway from College Park. He had passed away in 2004, but Tim was able to find the name and number for Nininger's widow, Eleanor. After a couple of attempts at convincing the nearly hundred-year-old woman that he was not a telemarketer, Tim managed to confirm that the cube had indeed belonged to Robert Nininger, and that Eleanor had donated it, along with a large portion of his rock collection, to one of her husband's colleagues at the University of Maryland after his death. Placed in an office drawer, it had then been forgotten.

The son of Harvey Nininger, a famous geologist who pioneered the study of meteorites, Nininger had followed in his father's footsteps into geology. By the time he had written the book that Tim would stumble on, Nininger was working for the Atomic Energy Commission as the deputy assistant director for exploration—a division of the AEC devoted to finding new sources of uranium and developing better methods of radioactive prospecting. Years earlier, Nininger had been working for U.S. Vanadium, a subsidiary of Union Mines, when he was drafted into the army. During the early years of the war he had served as a field geologist in the Strategic Minerals Program, but soon he found himself drafted into the uranium procurement effort of the Manhattan Project.

Nininger's name would appear a second time, months later, as the search for the history of the mysterious cubes continued: in small print on a large, folded piece of crackling brown paper that depicted the organizational chart for the Murray Hill area of the Manhattan Engineer District.

30 | THE RACE

WHEN NEWS OF THE DISCOVERY of fission made its way across the Atlantic to the United States in 1939, the American physicists, and the European scientists who had made the United States their new home, recognized the discovery as a major inflection point in the course of human history. Many hoped that unlocking the power stored within the atomic nucleus would provide a path to supplying humanity's ever-increasing demand for energy, as resources of fossil fuels were already beginning to diminish. Atomic energy, they hoped, could be a gift to humanity guiding us toward a gleaming future.

But what was also apparent to those reading Meitner and Frisch's report was the very real possibility of this discovery being used to build a new weapon more powerful than any instrument of war previously devised by man. Particularly for the scientists who had escaped from the wave of fascism sweeping Europe, the stakes of this possibility were too high to ignore. Arthur Compton, a well-known physicist and one of the early Manhattan Project leaders, explained that "there was every reason to believe that the Nazis saw in the atomic bomb the possibility of a new weapon of decisive importance. If they succeeded first they would have in their hands control of the world."

Despite the alarm of many American scientists, work on the American nuclear program began slowly. It was unclear to all involved in the early days of the work if the production of a nuclear weapon would even be feasible. Initial results from Princeton and Columbia, confirming and following up on Meitner's and Hahn's conclusions, indicated that a chain reaction of uranium fissions was likely, but there was still much that remained unknown about this process: how much energy it might release; if the explosion, once achieved, could be contained or controlled; and how much material might be required to make such a weapon possible.

In the months after the presentation of Einstein's famous letter to President Roosevelt warning of the possibility of the Germans using fission as a weapon, support from the US government in the form of the Uranium Committee was more symbolic than practical. In the first year, only about $6,000 of government funding were put toward the project, the rest being supplied by the home universities and institutions of the small community of scientists who had begun to work on the problem.

The first step toward determining any of the necessary answers these scientists sought lay in trying to create a self-sustaining nuclear chain reaction. Evidence pointed toward the minority isotope of uranium, ^{235}U, as the isotope actively involved in fission. By the summer of 1939, Enrico Fermi and his new Columbia associates, working in parallel with Eugene Wigner and Henry Smyth at Princeton, had determined that producing a nuclear chain reaction would require either isolating this fissile isotope in larger proportion than found in nature, or devising a way to slow down incoming neutrons to allow the ^{235}U to capture them, or perhaps some combination of both.

While early results, and speculation about their meaning, began to fly around the scientific community, none of this information reached the typical channels of science communication. Urged by Leo Szilard in the earliest days of the project, the US-based scientists had all agreed to hold back from official publication any and all results that might be of potential military importance. The journals and news outlets fell silent on the matter of nuclear fission. This did not mean, however, that communication between scientists at their respective universities had also ceased. Instead, the scientists hit the rails, ping-ponging across the country by train to consult with one another as each new detail of the problem was revealed.

By late 1941 it was increasingly clear that the United States' entrance into the war was only a matter of time. Scientists were being mobilized across the country to aid in war-related research. For the physicists working on fission, the decisive moment was at hand. It was clear that in order to accomplish the enormous task of building a nuclear weapon, an immense amount of financial and logistical support from the government would be required.

On a chilly Chicago evening in September 1941, Arthur Compton, who had recently been placed at the helm of the committee of the National Academy of

Sciences that was tasked with advising the government on the atomic program, arranged a meeting at his home with Harvard president James B. Conant and physicist and cyclotron builder Ernest Lawrence. Conant had been appointed the new chair of the National Defense Research Committee, replacing Vannevar Bush, who was now heading the larger, overarching Office of Scientific Research and Development; if additional support was to be obtained from the US government, Conant would have to make that request. But the Cambridge chemist was hesitant. All the early reports on the atomic program seemed to indicate that the production of a functional nuclear weapon would require years and unprecedented resources and funding. He felt that these resources, not to mention the scientists themselves, might better serve the war effort by being allocated to other necessary tasks.

Over coffee, sitting around the fireplace in the living room of Compton's house that evening, his wife, Betty, puttering around on the floor above them, Lawrence and Compton attempted to change Conant's mind. While the results did indeed indicate that the project would take immense effort, they also indicated, Compton argued, that an atomic weapon was actually feasible. And if it was possible to build such a weapon, there was no reason to suspect that the Germans, with their own scientific resources and two-year head start aided by the immense Nazi war machine, were not making a serious play for nuclear power. By the end of the evening, Conant was convinced. But, he warned, the path they were headed on would be long and difficult. "If such a weapon is going to be made, we must do it," Conant said. "We can't afford not to. But I'm here to tell you, nothing significant will happen on such a job as this unless we get into it with everything we've got."

Were Lawrence and Compton willing, Conant asked, to commit their every waking moment, possibly for several years, to seeing through this goal? They answered in the affirmative, as would many of their colleagues over the coming months. "For us the fight had now begun," Compton wrote of that evening. "No effort on our part was to be spared. But the full strength of the nation must be rallied. Nothing less would be enough."

Given the green light to pursue the development of a nuclear weapon without delay, Compton, Lawrence, Fermi, and their colleagues began to map the path forward. As Fermi began work on constructing an atomic pile, having moved his work from Columbia to the University of Chicago, the conversations began to shift toward what would happen after this important proof of principle of criticality was achieved.

As the scientists saw it, there were two paths toward making an atomic explosive. One was to devise a way to separate out the minority isotope of uranium, ^{235}U, from the noncontributing ^{238}U, and consolidate it in a highly enriched form. Accomplishing this task was complex. The two isotopes are nearly identical, ruling out any chemically based extraction methods—only the slight weight difference between the two atoms could be exploited.

Four possible methods of separation were proposed, each with varied pros and cons. Thermal diffusion relies on the tendency of the lighter uranium atoms to move toward a source of heat while the heavier isotope tends to concentrate in a colder area. The National Bureau of Standards was looking into this possibility, but the cost was seemingly prohibitive. Gaseous diffusion involves vaporizing a uranium compound and forcing it through a porous barrier; the lighter isotopes will reach the other side of the barrier first. This process had shown some promising initial results at Columbia University but was still under development. The centrifuge method, the same method favored by Paul Harteck in Germany, was being studied at the University of Virginia.

Finally, electromagnetic separation was being pursued in its early form by Lawrence and his team at Berkeley with their cyclotron. With the scientists using the immense particle accelerator Lawrence was building in his new laboratory on top of the hill at Berkeley, the two isotopes would be separated based on the slightly different circular path of an accelerated ^{238}U, compared to its lighter counterpart. This method was efficient for producing very pure ^{235}U but required a great deal of time and energy to produce meaningful quantities.

There was also another option to consider for producing a nuclear explosive: plutonium. While the majority isotope of uranium, ^{238}U, does not chain react, it is also not completely useless. The scientists postulated, correctly, that when subjected to a steady nuclear chain reaction, an atom of ^{238}U can absorb one of the free neutrons flying around the core. With the additional neutron, the uranium atom becomes the extremely unstable isotope ^{239}U, which immediately undergoes two consecutive beta decays* to become, first, neptunium-239, and then in short order plutonium-239.

Plutonium, element 94 on the periodic table, does not occur naturally, but the possibility of manufacturing it in a lab using a nuclear chain reaction was surmised shortly after the element's initial discovery at Berkeley in

---

* Beta decay involves the loss of an electron and an antineutrino.

1940. It was thought that this new element, like ^{235}U, would be fissile—able to participate in a chain reaction and potentially be used as the core of a bomb. A reactor, built from natural, unenriched uranium, could be used to generate this new element inside its core. As plutonium is chemically distinct from uranium, separating out the newly formed element could be accomplished with already-understood chemical methods. It was estimated that up to a pound of the new element could theoretically be made and extracted using this method every four days in an industrial-scale reactor.

With the additional possibilities presented by plutonium, the Manhattan Project scientists now had five feasible methods of producing not one but two types of nuclear explosive. Rather than pick one single method to pursue, the scientists instead tackled all methods at once, sharing information with each other that was learned along the way. "Though in one sense competing," Compton later wrote, "these various groups were also co-operating closely." The pursuit of these multiple possibilities in parallel greatly inflated the overall cost of the project, to its eventual total of over $2.2 billion.*

Eventually, after the US Army and General Leslie Groves formally took over the project, the isotope separation work would be moved to the new Manhattan Project site at Oak Ridge, Tennessee, while plutonium manufacturing became the purview of the facilities in Hanford, Washington. Operations at both sites worked tirelessly to supply the scientists at the third Manhattan Project site, Los Alamos, with the fissile material they needed to carry out their experiments, and eventually, to assemble the first atomic weapons. In August 1945, when the two bombs were dropped on Japan, one contained a core of ^{235}U, while the other contained plutonium.

Before any of this work could take place, the Manhattan Project would first need to obtain adequate amounts of uranium to supply both enrichment and plutonium production. Henry Smyth, who summarized the work of the Manhattan Project in a thin volume that was shared with the stunned world in the aftermath of Hiroshima and Nagasaki, concisely stated the challenge that lay before the scientists in 1941: "Obviously there would be no point in undertaking this whole project if it were not going to be possible to find enough uranium for producing the bombs."

---

* About $33 billion today.

31 | BELGIAN URANIUM

FROM THE TIME OF its initial discovery in 1789, until Meitner and Hahn's discovery of fission in 1939, uranium was relatively useless. Small amounts were found to be effective in creating brightly colored glass and pottery, but there were few other practical uses for the black stones dredged up from the silver mines in the Bohemian mountains. In fact, most uranium ore was so worthless in the late 1800s and early 1900s that when Marie and Pierre Curie requested a shipment of pitchblende from the Joachimsthal mines, they were able to obtain nearly eleven tons of the black rock for little more than the cost of transport.

Even the Curies were not really interested in the uranium content that was present in the rocks that were piled high outside their Paris lab. Instead, they knew from the residual radioactive measurements that remained in the mineral compound after all the uranium had been removed that these rocks contained trace amounts of something else—something new. Through backbreaking work over many months, the Curies painstakingly extracted from that mountain of radioactive rubble a paltry few grams of a new element: radium.

The discovery quickly captured the public's imagination. Radium was found to be useful in treating cancer, and its novel properties, particularly its luminous glow when mixed with zinc sulfide, gave way to dozens of industrial applications, including, most famously, for painting watch faces that would glow in the dark.* Hailed as a magical cure for all ailments, radium found its way into commercial products—from toothpaste to children's toys to quack

---

* Slowly poisoning the women paid to paint the dials.

medical devices to chocolate. For a time, it was even widely believed that a daily dose of the radioactive element was beneficial for one's health, and a ritual of drinking water from a ceramic crock that had been lined internally with a radium-laced coating became a wellness fad. While the devastating health effects of radium consumption would soon become clear, for a time, demand for the precious material skyrocketed, at one point exceeding $100,000 per gram, making it the most expensive element on the planet.

Producing radium, however, is extremely difficult. Radium does not exist on its own anywhere on Earth. Producing it requires painstaking extraction from uranium ore, as was done by the Curies. But radium is such a small component in these ores that often several tons of uranium ore must be processed in order to produce just a gram or two. And so, as demand for radium exploded, demand for uranium ore, and lots of it, grew exponentially in tandem.

In the early 1900s there were few known sources of uranium ore on the planet. The Joachimsthal mines in Bavaria, where the Curies had sourced their pitchblende, began actively producing uranium in addition to silver. In 1915 a far higher grade of uranium ore was discovered in the Shinkolobwe mine in what was then the Belgian Congo in Africa.

Belgian mining in what is today the Democratic Republic of the Congo began in 1800 under the rule of Belgium's King Leopold II. Unlike in Belgium itself, where Leopold served as the constitutional monarch, in the Congo he owned outright the entirety of what was then called the Congo Free State. With no oversight from the Belgian government, Leopold ruled the territory through violence and terror. In a pamphlet entitled *The Crime of the Congo*, Sir Arthur Conan Doyle wrote, "There is not a grotesque, obscene or ferocious torture which human ingenuity could invent which has not been used against these harmless and helpless people." The inhabitants of the Congo basin were subjected to unspeakable cruelties as they were forced to strip the land of is most valuable resources including ivory, rubber, and substantial mineral deposits for export to Europe.

Eventually, international pressure began to mount on both sides of the Atlantic for the Belgian government to act, with writers like Mark Twain also adding their voices to the opposition:

The royal palace of Belgium is still what it has been for fourteen years—the den of a wild beast—King Leopold II—who for money's sake mutilates, murders, and starves half a million of friendless and helpless poor natives in the Congo State every year, and does it by the silent consent of all the Christian powers.

In 1908 the pressure finally succeeded in forcing the Belgian government to annex the territory officially as a colony. It was hoped this move would help to rein in some of the most egregious cruelties being perpetrated, though for the Congolese under the yoke of essential slavery, the change in status did little to improve their conditions.

What the annexation did alter, however, was access of private companies to the resources of the African territory. European corporations could now bring their own operations into the area in order to profit off the land and people. One such company was formed under the name Union Minière du Haut-Katanga (Mining Union of Upper Katanga), which, with the help of the Belgian colonial state, gained control of 7,700 square miles of land that they began strip-mining for lucrative deposits of copper—their primary export—along with bismuth, cobalt, tin, and zinc.

As throughout the rest of the Congo, conditions at the Union Minière mines were deplorable—the miners were paid next to nothing for their dangerous and backbreaking labor. Disease and malnutrition ran rampant through the mining camp populations, while impossibly high tax burdens ensured that few who entered the mines were able to ever escape alive. One in ten miners died each year.

While the copper deposits provided more than enough profit for Union Minière to operate, it wasn't until 1915, when uranium ore was discovered at their mine at Shinkolobwe, that the true value of the Congo soil was realized. A high-grade uranium ore—in fact, the highest-grade uranium ore anywhere in the world, often measured at between 65 and 75 percent pure uranium—was found beneath the ground, and soon miners were hauling tons of black stones from the earth. The best of these rocks—those most likely to contain larger amounts of radium—were taken by train and then by steamboat to Belgium, where they were crushed, and the powder was stored in barrels inside the Union Minière warehouses to await processing and extraction.

The radium boom had died down significantly by the 1930s, and by 1937, Union Minière had amassed a large enough stockpile of uranium ore to supply the global uranium and radium markets for decades. With a massive glut of material safely stashed in warehouses, mining operations at Shinkolobwe were stopped when World War II broke out. In 1940, when Nazi forces invaded Germany, the huge stockpiles of Union Minière's uranium ore were still sitting in the warehouses in Olen, outside Brussels. It was these stores of uranium that would fuel Germany's nuclear research program.

But unbeknownst to everyone on both sides of the Atlantic, the US nuclear program would also be fueled largely by uranium from the Congo. In the days just after the German invasion of Belgium, whispered rumors began to fly about a ship packed with uranium ore that was making its way across the ocean. These rumors remained unsubstantiated for months until 1942, when Lieutenant Colonel Kenneth Nichols, the newly appointed head of the Manhattan Engineer District, the administrative arm of the Manhattan Project, asked for a meeting with Edgar Sengier, the head of Union Minière, who had escaped the invasion and was living and operating the company out of New York City. While Nichols knew that the mines at Shinkolobwe had recently been flooded and were inoperable, he hoped to talk to the Belgian about reopening the mine and the eventual purchase and shipment of Congolese uranium to the United States. When Nichols arrived at Sengier's office, he was surprised to find that Sengier not only already knew why he had come but had been expecting someone from the US government to find him for some time. Sengier explained that Union Minière would be happy to sell their substantial supplies of uranium ore to the United States and that these supplies were much closer at hand than Nichols had thought.

"I've been waiting for you to come," Sengier said to Nichols. "I notified your government a year ago that I had laid down twelve hundred tons of uranium ore on the wharves here in New York. If you now want it, here it is."

In the last days before the invasion, Sengier explained, guessing that uranium was of military value, he had ordered the rumored ship to be filled with barrels of high-grade uranium ore. These rescued supplies had been sitting for months in a warehouse in Port Richmond on New York's Staten Island. Stunned, Nichols enthusiastically agreed to the immediate purchase of the material.

32 | THE COMBINED DEVELOPMENT TRUST

WITH THE ACQUISITION of the sequestered Belgian/Congolese uranium supplies, the Manhattan Project likely met their immediate uranium needs for building their nuclear weapons. But, while only three nuclear weapons were actually detonated during the course of World War II, production of these novel explosives was always expected to extend far beyond the immediate circumstances of the war. The facilities erected by General Groves and the Manhattan Project for creating the first bombs were not built as temporary wartime structures to provide a place of work toward a limited goal but were instead massive permanent installations intended to continue functioning beyond the war, for decades to come. Groves wasn't just building a singular weapon to end the war; he was building an arsenal.

Doing so would require access to much larger supplies of uranium than were available to the Manhattan Project at the outset of their work, so identifying potential other sources of the radioactive material quickly became a top priority. As Groves explained in his memoir, "We wanted to be certain that at the end of the war we would not find ourselves in the embarrassing position of having the plants, the knowledge and the skills, but no raw materials to work with."

Also looming large over the Manhattan Project's procurement efforts, as it did with most other aspects of the project, was the growing tension between the United States and the Soviet Union. It was well known that the Soviets were beginning to piece together their own nuclear weapons program, for which they would also need considerable supplies of uranium. Groves hoped that if the United States and its allies could quickly find and lay claim to as many significant sources of uranium around the globe as possible, they might corner the global market on fissile material and thereby deny these resources to the Soviets.

What was not yet understood at the time was just how ubiquitous uranium is within the earth's crust—it can be found in various concentrations on every continent. As Richard Rhodes put it in his detailed account of the Manhattan Project, "He might as well have tried to hoard the sea." But Groves was certainly determined to try.

In the search for additional uranium, Groves and his team started at home. While nothing as high grade as the ore from the Congo could be found in North America, less-concentrated sources of uranium ore were known to exist. A Canadian firm, the Eldorado Mining Company, had its own stockpiled supply of uranium ore from a mine near Great Bear Lake, just outside the Arctic Circle. In 1941 Eldorado had accepted a contract from the Office of Scientific Research and Development to sell these supplies and to continue producing uranium for the United States.

Beyond securing North American uranium, Groves also hoped to gain sole access to all future uranium coming from the Congo. Sengier and Union Minière had been all too willing to arrange for the United States to purchase the stocks of material that the company had stashed in New York, but Groves hoped to be able to establish a larger agreement for the continued supply of material from Africa after the war. One of Colonel Lansdale's colleagues from his Ohio law firm, who happened to be fluent in French, was brought into the project to negotiate a potential deal with Sengier and the other exiled Belgians. But, as Sengier reminded them, the mines at Shinkolobwe had been flooded just before the war and had been rendered inoperable. The Union Minière officials showed little interest in discussions to repair and reopen them.

The Belgians' recalcitrance forced Groves to ask for assistance from his counterparts in the British atomic program, Tube Alloys. Not only did the British own a significant minority stake in Union Minière but the exiled Belgian government was being hosted in London, giving the British more bargaining power to leverage against the Belgian company than the Americans had. But the British wanted something in return for their help: access to the uranium supplies that the Manhattan Project was amassing. Much of Britain's Tube Alloys project had been moved from England to the United States by the security-obsessed Groves, and many scientists in the United Kingdom had been left feeling locked out of atomic research. The Belgian problem presented them with an opportunity to regain a little control. A proposal was made to establish an official agreement between the United States and the United Kingdom to jointly endeavor to find, hold, and—most important for the British—share all possible global sources of uranium around the world, including the Belgian Congo.

The plan was agreed to, and negotiations of the terms for this top-secret arrangement began in the spring of 1943 between Britain's Sir John Anderson, the chancellor of the exchequer, and the US ambassador to England, John G. Winant. In an unorthodox turn of events, President Roosevelt's secretary of state, Cordell Hull, to whom Ambassador Winant usually answered, was kept entirely in the dark about these discussions. Instead, the ambassador took his direction from the US secretary of war, Henry Stimson. Asked later about how he felt about excluding his boss from these talks, Winant replied that he had felt no conflict at all—that his job was to represent the president, not the secretary.

The terms took several weeks to iron out. Following advice from Colonel Lansdale, the agreement was forged as a common law trust established between the two nations in an effort to avoid congressional oversight. The new trust, it was agreed, ought to include Canada, and the Commonwealth was given one of the six positions on the board of directors. It was also agreed that thorium would be included as well as uranium in the procurement efforts. Winston Churchill signed the final agreement of the Combined Development Trust on June 1, 1943, with Roosevelt adding his signature when the copies arrived in the United States on June 13.

33 | MURRAY HILL

AGREEING TO SHARE CONTROL of as much uranium as possible was one thing, but actually identifying sources of uranium around the globe and then obtaining and processing the material was another.

In his memoir of the Manhattan Project, Arthur Compton wrote that "the story of the supply of uranium is by itself, a thrilling one, and the production of enough pure metallic uranium to do our task in time was a technological and industrial miracle." The establishment of the Combined Development Trust had broadly set out lofty and long-term goals for uranium procurement, but the actual work of obtaining and processing uranium, as well as the many other materials necessary for building a nuclear weapon, would require a massive administrative and industrial effort.

With this in mind, on February 5, 1943, Groves presented a plan to the Military Policy Committee for the institution of a new area within the Manhattan Project. The Madison Square area, as it became known, would be broadly responsible for obtaining and processing all the uranium and other materials required by the project. The committee agreed and Lieutenant Colonel John R. Ruhoff was placed in charge as area engineer.

While few appreciable sources of uranium were known prior to the war, it was suspected that other significant deposits might exist around the globe, waiting to be found. Locating these sources would require a complete review of all existing geology publications, surveys of locations of interest, and, most important, the expertise of geologists.

Rather than build up the infrastructure for this project from scratch, Groves and his team looked to their industry contractors for help in getting started.

221

Union Mines, a major US-based mining corporation, already had a division devoted to geological surveys and exploration, and its status as a large international corporation that was interested in a variety of metals, including uranium, made their involvement in the Manhattan Project's work likely to go largely unnoticed. They agreed to take on the assignment.

The project was given the code name Murray Hill, after the New York neighborhood where Union Mines set up a small office as an operating base, and a specialized subarea under Madison Square was established to oversee the work. To lead this new area, Groves wrote in his memoir, he was looking for "a man who was experienced in the oil industry . . . [who] would be used to making quick, conclusive decisions, based, if necessary, on very limited information. I did not want anyone who would always insist on 100 per cent proof before making a move." A search of over a million officer records produced the name of US Army Corps of Engineers Major Paul L. Guarin.

Originally from Harrisburg, Pennsylvania, Guarin was a mechanical engineering student working his way through Penn State when the United States entered World War I. After his best friend from childhood was killed on the front lines in Europe, Guarin, who was still too young to enlist, became determined to join the war effort. But his mother refused to "fib" on his application, and by the time he was able to talk his grandparents into signing doctored paperwork featuring an incorrect birth date, the war was coming to its end. So instead of heading to the front in Europe, Guarin returned to Penn State and finished his degree. After graduation, hoping to avoid a life in the smelting factories in central Pennsylvania, Guarin boarded a freighter headed south. In exchange for passage, he worked as a second cook until the vessel docked in Galveston, Texas. From there, he made his way to Houston, eventually finding his way into the oil industry.

When World War II broke out, Guarin was working for Shell Oil and living in Oklahoma with his wife and two children. Friends and family laughed when Guarin, now nearly forty, and who by this time suffered from a stomach ulcer and was nearly blind in his left eye, announced that he would be joining this new war effort. But Guarin was not going to miss this second chance to serve. With his doctor, he came up with a plan to pass the physical exams required for entry—he would go to the exam with a stomach full of oatmeal to hide the ulcer and would memorize the vision chart before being faking the exam with his bad eye. Their plan somehow worked, and Guarin enlisted in the US

Army Corps of Engineers. He was working to build POW camps for captured Germans outside Huntsville, Texas, when he got a call from General Groves's.

Guarin flew to Washington a few days later and made his way to Groves's office on the fifth floor of the new War Department building. The general's ever-present secretary, Jean O'Leary, sat at her desk outside the office front door, and two other secretaries were working away in the next room. When Guarin was shown into Groves's office, Groves rose from his chair and towered over Guarin, as he did over pretty much everyone else, as he shook the Texan's hand.

"The president has tasked the Corps," Groves cryptically explained as they sat down, "with producing something in the next year or two that is going to be impossible to accomplish. That is all I can tell you, except that the project is all volunteer."

Guarin was instantly intrigued and didn't hesitate in his reply: "Sign me up!"

He was promoted to lieutenant colonel, and within a few weeks, Guarin had moved his family from Dallas to the New York City suburb of Scarsdale. From his new office in Manhattan, Guarin began his work overseeing a massive global search for potential sources of uranium.

In addition to the team of researchers provided to the Manhattan Project by Union Mines, Guarin was assisted by civilian geologists George C. Selfridge, a professor at the University of Utah, and George W. Bain, a professor at Amherst College, as well as by mining engineer Frank J. Belina.

Work at the Murray Hill area was roughly divided into four objectives. Their primary task was extensive bibliographic research undertaken in the New York City headquarters. All available geological literature and reports in all languages were examined for any and all reported mentions of uranium around the world. In total, more than sixty thousand volumes were examined by this team in just a few months.

This literature research then informed Murray Hill's second objective: the work of the Field Exploration Division. Based on the identified areas of interest, Union Mines would dispatch small parties of geologists and mining engineers to search for evidence of uranium deposits. The Murray Hill team fanned out around the globe, as mines and geological formations in more than twenty foreign countries and in thirty-six states were included in these expeditions. While many of their ventures were fruitless, in some locations the geologists found precisely what they had been looking for. An investigation of active mines in Colorado revealed a significant supply of uranium that had been left in the tailings (discharged waste rock) outside vanadium mines. In all, Murray Hill was able to recover at least five hundred tons of uranium oxide from these waste piles. The deposit was so significant, in fact, that a secondary field office of the Murray Hill area was established in Grand Junction, Colorado, to support this part of the area's mission.

Identifying uranium deposits was, at the time, an inexact science, and better methods of prospecting and mineralogical research, including improved designs for portable Geiger counters, became the third focus of Murray Hill in the Exploration Research Division. And finally, developing methods of efficiently refining the many different types of uranium ores that were being discovered throughout the world were the focus of Murray Hill's Metallurgical Research Division.

While the Murray Hill area was small, especially when compared to other areas of the Manhattan Engineer District, only costing about $600,000 a year to run, their methods were effective. In a matter of months, hundreds of tons of uranium ore were identified in locations the world over.

One of the civilian geologists who was recruited to work on the project, Professor George Bain, showed "a remarkably thorough knowledge of geological formations around the world" and quickly developed a knack for locating overlooked sources of uranium, including extraction from oil and coal. In one instance, Bain happened to recall one afternoon that on a recent trip to South Africa he had found some uranium-bearing rocks in the area around the Witwatersrand gold mines. There was no evidence in the literature to support the presence of uranium in that region, but convinced that he was right, Bain boarded a train back to Amherst to retrieve a sample he had brought back from the site for his personal collection. Measurements of this small rock revealed a significant concentration of uranium. By 1959 this site had produced over $150 million worth of uranium ore.

In addition to the massive amounts of uranium that the geological sleuths of Murray Hill found around the globe, Lieutenant Colonel Paul Guarin and his team of geologists were also the recipients of another, unexpected source of uranium. When Colonel Boris Pash and his Alsos team discovered the remaining stashes of uranium ore in Belgium, and a few weeks later, in southern France, the material, which was placed under the administrative purview of the Combined Development Trust, became the responsibility of Murray Hill. The team handled and processed these shipments of raw ore from Europe in much the same way as they had for other ore shipments that were arriving from around the world.

One day a different kind of uranium shipment arrived. Instead of the small barrels that had been typical of previous caches of captured uranium, a handful of small but shockingly heavy crates were unloaded onto the pier. Newly appointed property accountability officer for the Murray Hill area, Robert Nininger, received the unusual shipment. The heavy lid was pried off one of the crates, revealing a collection of dozens of black cubes neatly stacked together inside their container. The geologists marveled at the dark objects—they had grown accustomed to seeing uranium in all its rocky mineral forms, but most had never seen it in its purified metallic state.

The box was soon resealed and sent on for processing—such a large amount of metallic uranium was too valuable to not be used—but not before a handful of the curious objects found their way into pockets, dark, heavy mementos of the buried treasure the Murray Hill geologists were so desperately seeking.

34 | MAKING METAL

URANIUM ORE, as it comes out of the ground, is a mixture of a small amount of radioactive material within other forms of minerals and rocks. Transforming raw ore into a usable format for use in the Manhattan Project required several labor-intensive steps.

The uranium first had to be separated from the useless materials that had formed alongside it. To accomplish this, the uranium-containing ore was pulverized into a powder before chemicals were added that, by reacting with the uranium, allowed it to be separated from the rest of the minerals in the rock.

After separation, the extracted uranium was in the form of a uranium compound powder, typically black oxide or yellow sodium diuranate. Further processing was then needed to purify this powder and then transform it into one of three primary feed material forms: uranium dioxide (UO_2), uranium hexa/tetrafluoride (UF_6/UF_4), or uranium metal.

Completing all these processes and delivering usable material to the major Manhattan Project sites was another task that fell under the responsibility of the Madison Square area and its seven subareas. While the geologists at Murray Hill, and its own subarea code-named Colorado, in Grand Junction, were busy researching and finding uranium supplies, the Iowa, St. Louis, Wilmington, Tonawanda, and Beverly areas oversaw its processing. Altogether, the work undertaken by the various branches of the Madison Square area accounted for a large share of the funding spent by the Manhattan Project, with $27 million spent on raw materials acquisition and another $58 million on refining and processing.

Most of the chemical processes for refining and converting uranium were still extremely new and being developed on the fly at the different Manhattan Engineer District sites. The most difficult of these materials to make turned out to be uranium metal.

Like their German counterparts, the Manhattan Project scientists knew that the higher density of uranium metal made it the ideal format for use in the construction of early nuclear reactor experiments. As early as December 1941 the scientists were already clamoring to get their hands on metallic uranium. Compton remembered one of these men pleading with him in his office: "All I need is a lump of uranium as big as this," indicating a ball the size of a grapefruit.

But even this small amount of metallic uranium was still out of reach. Before the war, Westinghouse Electric & Manufacturing Company had developed a method for producing small amounts of metallic uranium, mostly for laboratory measurement and novelty purposes. Each batch of metal produced by their process, which utilized potassium fluoride and sunlight, resulted in a lump of metal about the size of a quarter and cost about twenty dollars a gram. Westinghouse's head of research, Harvey Rentschler, made every effort to scale up production, including installing huge vats on the roof of their Bloomfield, New Jersey, facility, to hold the uranium solution up to the sun. But even under ideal weather conditions, Westinghouse reported, no more than a ton of uranium metal per month could realistically be expected. Producing enough material to supply both the uranium enrichment and plutonium production programs through this method alone would have taken years, and "to us," Compton recalled, "every week was important."

A potential alternative production method for metallic uranium was presented by Metal Hydrides, a small chemical firm operating out of Marblehead, Massachusetts. Founded by chemist Peter Alexander, Metal Hydrides specialized in manufacturing various metals using a calcium hydride reduction process. In 1941 it was determined that the calcium hydride process would also work for making uranium metal, and the company was offered a contract by the Office of Scientific Research and Development to manufacture metallic uranium for ten dollars per pound.

In order to meet this demand, Alexander set up a new manufacturing plant in the town of Beverly, Massachusetts. The plant was constructed on a plot of land on the bank of the Danvers River. The site was chosen in part for

An aerial view of Metal Hydrides on the Beverly waterfront, 1974. *Salem News / Beverly Historical Society*

its proximity to the train tracks that crossed the bridge connecting Beverly to its infamous neighbor, Salem, allowing easy access to incoming supplies. An estimated 107 men from the small coastal New England town were hired to work in the refinery, operating around the clock on a three-shift schedule. Within a few months, Metal Hydrides was churning out pounds of a fine uranium metal powder every day.

The format of the final product from Metal Hydrides presented the Manhattan Project scientists with the same difficulties that the Germans faced with their own metallic uranium powder; the uranium dust was both difficult to melt and cast and highly pyrophoric—barrels of it would routinely spontaneously catch fire. Since there was no easy way to put them out, these blazes would simply be left to burn themselves out in the yard.

But an even bigger roadblock was encountered just a few weeks after production at Metal Hydrides began. It was discovered that the calcium hydride being used in the metal reduction process contained a small amount of boron, and this trace contaminate was making its way into the resulting metal. From their experiments with graphite, the Manhattan Project scientists fully appreciated that even a small amount of boron, a powerful neutron absorber, in the uranium reactor fuel would be enough to halt a nuclear chain reaction. Production was immediately stopped, and it would take several months before experiments at the National Bureau of Standards and the Union Carbide & Carbon Corporation would lead to a process for producing a suitably pure calcium compound, allowing production to resume.

Work at Westinghouse had also stalled, and as a result, no significant amounts of usable metallic uranium were produced in the United States until the summer of 1942. The lack of uranium metal slowed the construction of Fermi's CP-1 reactor experiment and is in large part why the majority of the uranium elements used in that first reactor were made of pressed uranium oxide, instead of the originally intended metallic elements.

While Metal Hydrides and Westinghouse were fumbling, the Manhattan Project scientists began to develop a more scalable method of producing uranium metal in a nonpowder form. The solution they arrived at was controversial. It involved dissolving heated uranium nitrate in cooled ethyl ether. Ethyl ether is an unpredictable chemical to work with. It can react with oxygen in the air to form unstable peroxide compounds, which when disturbed by sudden movement or changes in temperature can violently explode.

Finding a corporation that was willing to take on this production method was difficult. Most of the industrial and chemical manufacturers had already taken on other wartime activities, and the unstable and explosive nature of the process did not provide any additional incentive. Eventually, Arthur Compton identified the Mallinckrodt Corporation out of St. Louis as a possible solution. The small chemical firm already specialized in the production of ether and other pure chemicals.

Compton traveled to St. Louis to meet with the company's owner, George Mallinckrodt, in his ornate family home near the city's central Forest Park. Together, they discussed in detail the requirements of the project and the possible dangers of the proposed process. By the end of the day, Mallinckrodt had agreed to participate, and the small company quickly turned nearly its full focus toward the production of uranium compounds.

Like most of the auxiliary industrial sites that were supporting the work of the Manhattan Project, the men working at the site in St. Louis, by and large, did not understand the context of the work they were performing. A *St. Louis Post-Dispatch* article describes how "code names such as Biscuit, Juice, Oats,

Cocoa, and Vitamin were slapped on all the steps of the process. Correspondence about the project read like a breakfast menu." Despite the veil of secrecy and a number of accidental explosions, work continued at a breakneck pace.

By 1943 Mallinckrodt was receiving regular shipments of uranium ore and processed powder and was successfully converting these supplies into the requisite uranium metal that was then sent on to supply both the enrichment facilities at Oak Ridge and the reactors at Hanford, as they worked to produce enough fissile material to build a bomb.

35 | THE LAST STOP

BY 1943 THE PRODUCTION of uranium metal had been transferred to Mallinckrodt in St. Louis and operations at Metal Hydrides in Massachusetts shifted from metal production to recovering and reprocessing scrap and by-product uranium from the other Manhattan Engineer District facilities. Sixteen men worked on the scrap recasting operation, which at its peak reprocessed about three thousand pounds of metal daily.

The majority of this work, like all other industrial areas of the Manhattan Engineer District, was conducted by Metal Hydrides employees and civilians, but the area itself was overseen by a General Groves–appointed area engineer. To oversee work at the Beverly area, Groves chose a young chemical engineer named Richard Duffey. Duffey was selected for the post, in part, because he held a patent for the production of calcium hydride—the vital step in the metal production process that had been used at the site.

Born in 1917 on a small farm in La Fontaine, Indiana, Duffey was an odd and enterprising boy who frequently trapped muskrats and other rodents for their fur, practiced roping on passing chickens, and after a visit to the circus, "rigged up a rope in the hay loft of the barn and taught himself tightrope walking." After graduating with his master's degree, Duffey was hired by Union Carbide in Buffalo, New York. When Union Carbide joined the war effort, Duffey was tapped to join their top-secret Manhattan Project work.

When the Murray Hill area received the shipment of confiscated German metal cubes in the summer of 1945, the material was scheduled to be processed like any other scrap uranium metal in the Manhattan Project. By this point in the race for nuclear energy, the United States had so far outstripped the efforts of their German counterparts that melting down and repurposing these cubes, which represented the full culmination of Werner Heisenberg's wartime nuclear research, fit easily within a literal day's work for the reprocessing team at Beverly.

The crates of cubes were loaded onto a train in New York and shipped north on what would become the last leg of their long journey. A few hours later, the train pulled up at the small station that still sits directly next to the lot that used to be the Metal Hydrides plant, underneath the Beverly–Salem bridge. The crates were unloaded, and carried inside, to be melted down and added to the United States' rapidly growing nuclear stockpile.

After the war, Duffey went back to school and completed his PhD at the University of Maryland, where he later become a professor and one of the founding directors of the university's small teaching nuclear reactor. Decades later, Tim Koeth would look through old samples in a cabinet tucked away behind that same reactor pool. There, sitting inside a tattered box on one of the green steel shelves, he found a small dark object: a cube of uranium metal that Duffey had plucked from the pile and saved from the furnaces at Beverly.

––––––––––

There are no records indicating the number of cubes transferred from Europe to Murray Hill, and even less information was preserved detailing the masses of uranium metal that were processed by Metal Hydrides. It is impossible to know how many of these artifacts were plucked off the pile in New York or Massachusetts as personal souvenirs.

Over the past few years, we have tried to track down as many of these remaining objects as possible. Most of the cubes that survive today are unofficial mementos, with no paperwork and only vaguely remembered family lore connecting them to their place in history. But with each discovery of a new cube, a little more has been added to the broader story.

One afternoon, several months after the first publication about our hunt for the uranium cubes in *Physics Today*, Tim opened his email to find a message from a Betsy Bloomer of Long Island, New York. The email read simply, "I have one of the missing children you are looking for," along with her phone number. Tim dialed the number immediately, and a soft, thin voice on the other end of the line answered, "That didn't take long."

Betsy, it turns out, is the daughter of Wilbur Kelley, the area engineer appointed to lead the Madison Square area in early 1945. He supervised both Murray Hill's search for uranium and the uranium scrap processing at Beverly. His diary from that year details several shipments of captured material from Europe, and while cubes are never explicitly mentioned, it seems certain that crates of the heavy black metal objects were among these deliveries. Kelley went on to spend his career working for the Atomic Energy Commission and amassed an impressive collection of nuclear artifacts, including one of the German uranium cubes, now stewarded by his daughter and her family.

36 | THE NEW URANIUM CLUB

OVER A THOUSAND CUBES of uranium metal were produced in Europe during World War II: 664 cubes that were used by Heisenberg in constructing B-VIII in his cavern laboratory in Haigerloch and an additional stash of about 400 or so cubes that Diebner had retained control of in Stadtilm. Today, only fourteen of these cubes are known to exist, the last remnants of Nazi Germany's failed attempt at nuclear supremacy.

Three-quarters of a century later, the remaining cubes have been scattered around the globe. They are now in the possession of a small group of individuals, a new sort of Uranium Club, who each through their own set of circumstances of history and fate have found themselves stewards of these unique relics.

Only a handful of cubes remain in modern-day Europe. In Paris, a cube left behind at Haigerloch by the Alsos Mission and later delivered to Frédéric Joliot-Curie now props open a door in the home of Marie Curie's granddaughter. Only three cubes remain in Germany. One is held by Germany's Federal Office of Radiation Protection. Another sits in its own custom display case among a collection of many other mineralogical and metallic curiosities in the Mineralogical Museum at the University of Bonn. And finally, Heisenberg's makeshift laboratory in the Black Forest has now become the Atomkeller Museum. Visitors to the site can see a cube, along with one of the one-centimeter-thick square plates, in a display case beside the pit in the cave floor where the reactor was once assembled. An aluminum replica of the full B-VIII core hangs above from the cave ceiling.

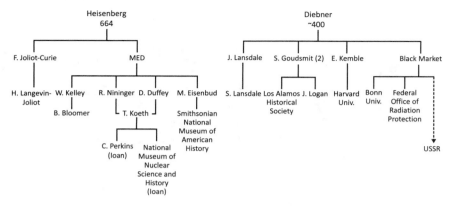

The new Uranium Club generational tree. *Created by author.*

The rest of the remaining cubes made their way to the United States. One of the two cubes that were taken by Samuel Goudsmit and displayed proudly on his desk throughout the rest of his career is now in the collection of the Los Alamos Historical Society in New Mexico. The other remains in a private collection. Goudsmit was not alone in returning to the States with his radioactive prizes. Two other cubes were taken directly out of Germany by Alsos Mission members: John Lansdale brought his own cube back, getting it cast in plastic before placing it in the window of his Ohio home office. And Edwin Kemble, the physics professor who had brought Goudsmit out to Harvard at the onset of the war, and who had spent a few weeks in Germany assisting during the final phases of the Alsos Mission, also brought a cube back over the Atlantic with him. Today, this object resides in a locked closet on an upper floor of the new Science Center, just north of Harvard Yard in Cambridge, Massachusetts. Black and heavy, this cube, which has had each of its corners shaved off, likely for sampling, is pulled out of its sheet-lead wrapping once a semester and passed around during introductory physics lectures, the unexpected heft of the object impressing even the most apathetic freshmen.

Another four cubes were saved from the scrap pile by members of the Murray Hill and Madison Square areas of the Manhattan Engineer District. In addition to the cubes taken by Robert Nininger and Richard Duffey, which both found their way to Tim Koeth, and the cube that once belonged to Wilbur Kelley, a fourth cube was donated by Kelley's colleague, Merril Eisenbud, to the Smithsonian Institution. This object is now part of the Medicine and Science

Collection at the Smithsonian's National Museum of American History. It sits in a crate, alongside a pseudosphere from CP-1, in one of the Smithsonian's vast storage facilities.

The path taken by the cube held in the collection at the Pacific Northwest National Laboratory in Washington State is far less clear. But in this case, science was able to make up for a lack of documentation. High-tech chemical analysis has allowed for the confirmation of its authenticity. The Rochester Institute of Technology also purportedly had a cube but disposed of it in the early 2000s, leaving no paper trail.

37 | SCIENCE IN CONTEXT

WHEN VIEWED AS A WHOLE, the final destination of each cube tells the complete story of their history from their manufacture in Germany, to their capture by US forces, and to their eventual repurposing for America's own nuclear ambitions. But in many ways the history of the cubes is much larger than the story of one set of ultimately failed experiments. Instead, this story offers insight into a pivotal moment in the history of science and the ways in which one group of scientists managed against great odds and under unfathomable pressure to harness the power of a new discovery, while another group failed at the same goal.

Though the differences between the US and German nuclear programs are innumerable, most of them can be distilled down to a fundamental difference in how science was viewed and practiced in the two countries. The German program approached the question of nuclear power on an academic scale, largely limiting its scope to the traditional boundaries of university research. And the physicists themselves, none more so than Heisenberg, treated the problem of building a nuclear reactor, as well as their own involvement in the program itself, as entirely separate and apart from the wider machinations and implications of the war happening around them. Experiments were carried out piecemeal in disparate laboratories, and proof of principle was sought at each step before any investment of time or resources was put toward what came next.

General Groves and the American program took the opposite approach. The US scientists worked tirelessly around the clock, pursuing every possible path forward with every resource at their disposal. While Groves, Oppenheimer, and Fermi arguably dominated the work of the Manhattan Project, the task could never have been accomplished without input from a massive team. Calculations, predictions, and results were checked and rechecked and discrepancies were thoroughly investigated. Unlike the German program, where the final word was often left to the opinion of one individual, in the United States, no one single voice was deemed unimpeachable. And the Americans also understood that harnessing the atom would take more than a village; it would take a whole nation. By the end of the war, there wasn't a city, university, or major corporation in America that hadn't in some way become a small cog in the massive machine of the Manhattan Project. As Niels Bohr would later observe, Groves had turned the whole country into a factory.

None of the work of the Manhattan Project would have been possible without the support of the US government. President Roosevelt and his administration believed in and supported the small group of scientists who came to them with a plan to accomplish the impossible, providing the program with the vast financial resources and administrative latitude that was needed to complete the task at hand. Across the Atlantic, the Nazi regime, embroiled in an atmosphere of mysticism and superstition, and utterly convinced of the inevitability of their eventual victory, showed little interest in their own scientists or the development of a nuclear program until it was too late.

But the Nazi government did more than just fail to support or promote the work of their nuclear scientists. In the years before the war, the demagoguery, antisemitism, and racism of the fascist ideology infiltrated and handicapped German physics. The scientific community was gutted by the loss of so many of their Jewish colleagues, and by the outright rejection by many prominent German scientists of entire fields of scientific inquiry based solely on the Jewish identity of the thinkers from whom they originated. This left a generation of German scientists ill prepared to tackle the new frontiers of modern physics.

The Manhattan Project, in contrast, embraced the scientific refugees who fled fascism in Europe and took full advantage of the expertise and diversity of thought they brought with them. Good science, they recognized, can come from anyone, regardless of religion or nationality. One can only imagine what greater feats the Manhattan Project team might have accomplished had women and people of color been included in these conversations as well.

As the dust settled in the years after the war, the notion that the Germans had allowed themselves to fall so far behind the scientists working in America was almost unfathomable to many in the international scientific community. Goudsmit writes in his memoir that he routinely encountered fellow scientists who would simply refuse to believe that Alsos had really found the full summation of the German program, insisting instead that somewhere in a secret bunker lay the true results of Heisenberg's work. Even today, there are historians of science who promote similar narratives. But when the full context of the German approach to the nuclear problem is considered in comparison to the American effort, its failure to accomplish this monumental task hardly seems surprising.

The history of the cubes not only allows us a window through which it is possible to view the reasons behind the failure of the German nuclear program; it also provides lessons for scientists working on the biggest problems facing humanity today. As Goudsmit aptly summarized: "These same mistakes are the principal ones that we too can make if we are not on guard against them."

The cube that Tim Koeth received in the Maryland parking lot, and that started a years-long quest for answers, has since found a new home at the National Museum of Nuclear Science and History in Albuquerque, New Mexico. This small but mighty museum serves as the final home for many relics of the atomic arms race during World War II, and the race for nuclear armament that followed. The cube sits alone in a glass display case in the center of its own gallery, surrounded by panels describing its journey over the past eight decades. Visitors to the museum stare through the glass at its rough black surface, excited and a little bit nervous to find themselves so close to such a large piece of the powerful and almost mythical metal that has played such a large role in shaping the last century of human history.*

Around the corner from the cube's gallery, on an inconspicuous wall near the door leading out to the museum's outdoor display of bombs and bomber planes, hangs a small clock. Unlike a normal clock whose hands move

---

* The radiation levels around the cube in the exhibit are negligible.

continually with the passage of time, the time displayed on the face of this clock only changes when the world is faced with a new and dangerous crisis.

Started in 1947, the Doomsday Clock was developed in response to the immense death and destruction seen in Hiroshima and Nagasaki. During the tensions of the Cold War the Doomsday Clock was used as a method of describing how close the world was at any moment to a nuclear disaster. The closer the hands to "midnight," the more imminent the threat.

Though we are no longer in the throes of an explicit nuclear showdown, the clock is still in use today, and its hand positions are updated regularly by the members of the Bulletin of the Atomic Scientists. However, the time shown on the face now takes into account more than just the threat of nuclear war. Global catastrophe, as we have all collectively learned in recent years, may be carried on an intercontinental ballistic missile, but it is just as likely to come in the form of a virus carried on a cough, or in the crumbling of ancient ice.

For the scientists working in the United States to harness nuclear energy, the enormity of the threat posed by Nazi Germany was both clear and tangible. Today, the existential threats to humanity's existence are far less easily visualized, but no less dangerous. At the time of writing, SARS-CoV-2, the virus responsible for causing COVID-19, has taken the lives of roughly 6.7 million people around the world. During the course of the pandemic, science disbelief and denial of scientific analysis and conclusions found their way from the realm of large-scale public policy into living rooms and communities, as millions of people refused to follow medical and scientific advice on social distancing and mask wearing. These individuals, encouraged by misinformation and science skepticism streaming from the highest offices of the US government and an administration indifferent to data and facts, accelerated the spread of the lethal pathogen.

But it is not just pandemics that threaten humanity today. The Intergovernmental Panel on Climate Change warns that if global carbon emissions are not dramatically cut, we will soon face a dangerous future on an inhospitable Earth. The ten hottest years ever recorded by humans have all occurred within the last two decades, and over 99 percent of peer-reviewed scientific climate studies show that our global climate is changing as a direct result of human activity. The extreme weather fueled by climate change is now responsible for five million deaths a year globally. But despite these well-documented science and data-driven facts, climate change denial still runs rampant, particularly among the upper echelons of industry and government around the world.

The stakes for humanity have perhaps never been higher, but for both these crises—pandemics and climate change—science offers a pathway forward. The tireless work of biologists and doctors gave us a vaccine that has saved millions of lives and offered us a return to normal. And scientists and engineers are constantly working to develop zero-carbon technologies with which to power our world without poisoning our planet. If the last century was the century of the physicist, this century will be the domain of the biologist and the climate scientist.

As the history of the cubes illustrates, new discoveries developed in laboratories cannot alone alter the course of human history. Instead, there must be a united global collaborative effort within government, industry, and the population at large to implement them. Vaccines only work when people are willing to get them; carbon emissions will only fall when we all do our part to limit our own carbon footprint.

When first instituted in 1947, the Doomsday Clock was set at seven minutes to midnight. Today, humanity only has ninety seconds left. Just as harnessing the power of the atom defined and forever altered the history of humanity in the twentieth century, how we collectively rise to these new challenges, and move back the hands of the Doomsday Clock, or fail to do so, will define the next chapter of human existence.

EPILOGUE

IN THE SPRING OF 2019 Tim Koeth was invited to give a talk about the uranium cubes at the National Institute of Standards and Technology's Center for Neutron Research. This facility, which includes a nuclear reactor, is the site of complex and detailed research using neutrons produced by the reactor to probe materials and explore essential questions about the structure of the universe.

During his talk in the lecture hall tucked below the sprawling experimental floor in the reactor building, the scientists of NIST peppered Tim with questions: Did he think there were more cubes still in existence? Was there any way for the Nazi scientists to have successfully built a working reactor? What were the reasons for their ultimate failure?

As Tim answered these queries, many of which he had received before, a hand quietly went up in the back of the room.

"But what about the heavy water?" the inquiring audience member asked when Tim acknowledged his raised hand. "Do you know anything about where any of the water might have ended up?"

"No," replied Tim. "We have no idea where it might have gone."

The audience member's voice crackled with glee. "I do!" he said.

Uranium, of course, is only a part of what is required to build, or at least try to build, a nuclear reactor. A moderator is also essential. In addition to the buried cubes, the Alsos Mission members found and extricated several barrels of Germany's last remaining supplies of heavy water from where they had been stashed inside a nearby mill. This supply of precious liquid was presumed to have been moved to the States, along with the uranium, to be used in America's own nuclear program. But since barrels of water don't make good souvenirs or desk ornaments, their fate was largely presumed to be unknowable, and the water most likely lost.

When NIST's Health Physics Department had been preparing for a renovation of their building, they had come across three small aluminum dewars stashed under a sink. The labels taped to the front of the metal bottles indicated that the water they contained had been produced in the Vemork plant in Norway between 1942 and 1945 and had been used by the Germans in their reactor experiments.

Tags tied to the necks of the dewars provided another clue about the water's history. The small, tattered paper tied to each bottle read:

E. R. Smith

Nat. Bur. Stds.

April 29th, 1948

The NIST library keeps files on all former employees stretching back to the days when the institute was still called the National Bureau of Standards. A search revealed that E. R. Smith most likely refers to chemist Edgar Reynolds Smith, the first person to successfully extract a nearly pure sample of heavy water in 1931. His early experiment paved the way for Harold Urey's process in the months that followed. It would seem that a portion of the heavy water collected by the Alsos Mission was sent to Smith in the years after the war.

But unlike the uranium cubes, the jugs of heavy water were not kept at NIST as mementos or trinkets to be displayed. The label adhered to the barrel of the bottles gives some insight into the purpose their contents found at the institute. The water was kept not for its historical value but because of its "very very low tritium content," as is indicated on the bottle's label itself. Tritium, an isotope of hydrogen heavier than deuterium, was a by-product of the above-ground hydrogen bomb testing conducted by the United States and USSR during the early days of the Cold War. Today, as a result of these tests, it is virtually impossible to produce water or heavy water without trace tritium contamination, which can skew the results of some sensitive experiments. The water in the dewars, which was distilled before these tests, does not suffer from the same contamination. Throughout the 1970s and 1980s, this water was used repeatedly in experiments by NIST scientists who took advantage of its unique composition.

Today, one of the dewars sits in Tim's personal collection, reunited seventy years later and four thousand miles from its starting point with a uranium cube.

ACKNOWLEDGMENTS

MY ETERNAL THANKS are owed to Tim Koeth for introducing me to the amazing world of nuclear science and history and for bringing me along on such a spectacular adventure. I am certain this will not be the last project we work on, and I will forever be grateful to be a part of your "nuclear family." My gratitude also goes to my editor Jerry Pohlen and my agent Jessica Papin not only for their support and guidance, but also for their immediate enthusiasm for this unconventional story.

In researching this book I was immensely privileged to speak with many of the children of Alsos Mission and Manhattan Project members who were kind enough to share their families' stories with me, including Sally Lansdale, Chloe Pitard, Patricia Guarin-Roper, Patricia Walters, Jim Beatson, Kathy Fiebig, and especially Betsy Bloomer. I am also grateful to Mary Dorman and David Ebert for recognizing the significance of the cubes at UMD and for ensuring that they found their way to Tim. Thanks is also owed to Tim's wife Michelle, who has supported this project as she stoically does for all of Tim's fascinations.

None of the research and work for this project would have been possible without the support of so many people in the nuclear science and history community including Bruce Cameron Reed, Tim Gawney, Ray Smith, Tom Kunkle, Jon Schwantes, Britt Robertson, Roger Sherman, Carl Willis, Bill Kolb, and particularly the unparalleled Clay Perkins. Special thanks is also owed to Jim Walther and the entire team at the National Museum of Nuclear Science and History in Albuquerque, New Mexico, who have done such a wonderful job making a new home for the cube and its story.

I am also grateful to Dr. Robert Briber and Eric Chapman at the University of Maryland for their early support, which allowed me to pursue this project, and to Melissa Andreychek, who has helped us to effectively communicate this story from its very beginning. Thanks as well to all of Tim's colleagues and students, especially Noah Hopis, Patrick Park, Amber Johnson, Luke Gilde, and most especially Scott Moroch, for helping me and making space for me and this strange project in their research group.

I am deeply fortunate to have the support of my friends and family. Thank you to Audra Wormwald for reading an early draft of this book, and to Kaitlyn Cross and Alison Schuppert for listening to my endless explanations and angst. Thanks also to my father for supporting me, my brother for inspiring me, and especially my mother who has read, edited, and improved every piece of writing I have ever sent her way, including this book.

NOTES

3. A Brief History of Fission

"*our Marie Curie*": Ruth Lewis Sime, *Lise Meitner: A Life in Physics* (Berkeley: University of California Press, 1996), ix.

"*initial reaction was one of extreme*": Leslie R. Groves, *Now It Can Be Told: The Story of the Manhattan Project* (New York: De Capo Press, 1962), 4.

"*What little I knew*": Groves, 4.

"*You'll be interested to know*": Arthur Compton, *Atomic Quest: A Personal Narrative* (New York: Oxford University Press, 1956), 144.

"*Batter my heart*": Richard Rhodes, *The Making of the Atomic Bomb* (New York: Touchstone, 1986), 572.

"*told to lie face down*": Groves, *Now It Can Be Told*, 295.

4. The Lawyer: John Lansdale Jr.

"*After all . . . Dad lived*": Sally Lansdale, author interview, October 29, 2019.

"*had never really met*": Lansdale, author interview.

"*talent for fakery*": Maj. Montgomery C. Meigs, interview summary, conversation with John Lansdale, memorandum for record, March 1, 1982, Washington, DC.

"*Mother will never forgive*": Chloe Pitard, author interview, January 31, 2020.

"*BLIND ROAD—NO VISITORS*": John Lansdale, *John Lansdale, Jr: Military Service* (self-pub., 1987), 17.

"*trying to break the uranium atom*": Lansdale, 17.

"*one pound of U-235*": Lansdale, 21.

"*any connection between*": Lansdale, 19.

"*impress upon them*": Lansdale, 21.

"*He was in fact the only person*": Lansdale, 84.

5. The Soldier: Boris Pash

"a number of very intense": Lansdale, *Military Service*, 29.

"There was plenty of 'smoke'": Lansdale, 32.

"a rather aggressive search": Lansdale, 33.

"We were going to risk our necks": Boris Pash, *The Alsos Misson* (New York: Charter, 1969), 52.

6. Alsos in Italy

"While the major portion": George Strong to George Marshall, September 25, 1943, correspondence ("Top Secret") of the MED, 1942–1946, NARA II, RG77, M1109, roll 4.

"unique opportunity . . . [of] a dry run": Lansdale, *Military Service*, 41.

"Colonel . . . my buddy": Pash, *Alsos Misson*, 18.

"Well Gerry": Pash, 19.

"become substantially independent": "Non-Technical Report from Alsos Mission," January 20, 1944, correspondence ("Top Secret") of the MED, 1942–1946, NARA II, RG 77, M1109, roll 4.

"Sound intelligence is seldom served up": Pash, *Alsos Mission*, 29.

"the information obtained by the Mission": Vannevar Bush to Brig. Gen. Leslie Groves, re: Alsos, February 29, 1944, NARA II, RG 165, box 140, record copies.

7. The Scientist: Samuel Goudsmit

"My grandfather . . . was a tourist guide": S. A. Goudsmit, "The Discovery of Electron Spin" (lecture at Dutch Physical Society, April 1971), trans. J. H. van der Waals, http://www.physics.usu.edu/Wheeler/QuantumMechanics/QMGoudsmit.pdf.

"both looking thin, energetic": "A Scientist Views the Past: The Samuel A. Goudsmit Collection of Egyptian Antiquities," exhibit catalog, Kelsey Museum of Archeology, University of Michigan, Ann Arbor, January 30–May 9, 1982, 3.

"He told me, as we shook hands": Pash, *Alsos Mission*, 36.

"I promised that if any jumping": Pash, 36.

"The world has always admired the Germans": Samuel Goudsmit, *Alsos* (New York: Henry Schuman, 1947), 49.

8. Alsos in England

"no practical utilization": Thomas Powers, *Heisenberg's War* (New York: Alfred A. Knopf, 1993), 283.

"it was so clear": Major Montgomery C. Meigs, interview summary.

"But . . . it gave the handful": Groves, *Now It Can Be Told*, 199.

10. Paris

"never even saw the basement": Reginald Augustine, *A Great Big Glorious World: Adventures from Decatur to Casablanca to the ALSOS Mission, 1913–1945* (Kindle ed., 2018), 1786.

"resembled the prize floats": Pash, *Alsos Mission*, 64.

"There was feverish activity": Pash, 74.

"arrest and hold one Samuel Goudsmit": Pash, 74.

11. Belgium

"With the help of his raft": Pash, *Alsos Mission*, 79.

"We put on our Sunday": Goudsmit, *Alsos*, 59.

12. Unoccupied France

"The Pentagon kept firing": Pash, *Alsos Mission*, 100.

"Johnny, give special attention": Pash, 133.

"Water negative": Goudsmit, *Alsos*, 22.

"But . . . we never knew": Goudsmit, 30.

13. Strasbourg

"Don't you feel well?": Pash, *Alsos Mission*, 150.

"But he's a scientist": Pash, 156.

"We've got it!": Pash, 157.

"To an outsider, a professor is a professor": Goudsmit, *Alsos*, 32.

"biggest intelligence bombshell": Pash, *Alsos Mission*, 159.

"This is the most complete dependable": Memorandum for Maj. Gen. Clayton Bissel, from L. R. Groves, March 16, 1945, NARA II, RG 165, box 137, commendations.

14. Heidelberg

"they can work, scheme": Pash, *Alsos Mission*, 162.

"You know I am not the only scrounger": Pash, 167.

"filthy and . . . disgustingly unhygienic": Alsos Mission Diary (Germany) 1945 March–July, Boris T. Pash papers, box 2, folder 1, Hoover Institution Library and Archives.

15. Diebner's Lab

"worth even the sprained ankle": Goudsmit, *Alsos*, 96.

"an army ordinance car": Transcript of interview with Edwin Seaver on Alsos, 1947, Samuel A. Goudsmit papers, Series 01 Biographical Materials box 1, folder 4, Niels Bohr Library and Archives.

"What is this black stuff?": Goudsmit, *Alsos*, 88.

"merely reflected the pitiful smallness": Goudsmit, 89.

16. Operation Big

"most of the barrels": Groves, *Now It Can Be Told*, 237.

"the capture of this material": Pash, *Alsos Mission*, 199.

"Let rest be holy": Pash, 207.

"taking these towns more or less by surprise": Transcript of interview with Edwin Seaver on Alsos, 1947, Samuel A. Goudsmit papers, Series 01 Biographical Materials box 1, folder 4, Niels Bohr Library and Archives.

"excavating on the edge": Pash, 217.

"Boris Pash has hit the jackpot": Pash, 218.

17. Modern Physics

"Family lore has it": Werner Heisenberg and Elisabeth Heisenberg, *My Dear Li: Correspondence 1937–1946* (New Haven, CT: Yale University Press, 2011), 4.

"family of pedagogues": Elisabeth Heisenberg, *Inner Exile: Recollections of a Life with Werner Heisenberg* (Boston: Birkhäuser, 1984), 84.

"Always strive to be": Heisenberg, 85.

"He wrote [to Bohr]": Heisenberg, 61.

"I have been asked": W. Heisenberg and E. Heisenberg, *My Dear Li*, 65.

"Germany needs me": Goudsmit, *Alsos*, 112.

"At the time, he still firmly believed": Heisenberg, *Inner Exile*, 64.

"See Fermi, see Heisenberg": Fredric Alan Maxwell, "A Little Spying and Kidnapping Among Friends," *Michigan Today*, November 9, 2011.

"the conviction that we were sound asleep": Goudsmit, *Alsos*, 164.

18. Jewish Physics

"decadent Jewish spirit": Heisenberg, *Inner Exile*, 41.

"Before the war the Nazis had voiced": Goudsmit, *Alsos*, 141.

"Welteislehre or world ice theory": Goudsmit, 205.

"spectre of national hubris": Heisenberg, *Inner Exile*, 43.

"Heisenberg had underestimated": Heisenberg, 45.

"Have you already seen": Heisenberg, 47.

"*Einstein disciples*": "Weisse Judden in der Wissenschaft," *Das Schwarze Korps* 20, July 15, 1937, in *Physics and National Socialism: An Anthology of Primary Sources*, ed. Klaus Hentschel, trans. Ann M. Hentschel (Basel, Switzerland: Birkhäuser Verlag, 1996), 156.

"*Heisenberg is only one*": "Weisse Judden in der Wissenschaft," in *Physics and National Socialism*, 157.

"*Heisenberg was personally much too involved*": Heisenberg, *Inner Exile*, 61.

"*Heisenberg [was] a man of ideals*": Goudsmit, *Alsos*, 166.

"*One day . . . the Hitler regime will collapse*": Goudsmit, 114.

"*believes that Heisenberg is*": Heinrich Himmler, "Letter to Reinhard Heydrich [July 21, 1938]," in *Physics and National Socialism*, 175.

"*PS. I do find it appropriate*": Heinrich Himmler, "Letter to Werner Heisenberg [July 21, 1938]," in *Physics and National Socialism*, 177.

19. The Uranium Club

"*We take the liberty*": Rhodes, *Making of the Atomic Bomb*, 296.

20. How to Build a Nuclear Reactor

"*the entire radiation energy*": Jeremy Bernstein, *Hitler's Uranium Club: The Secret Recordings at Farm Hall*, 2nd ed. (New York: Copernicus, 2001), xxiv.

21. Early German Experiments

"*So overall I am really busy*": W. Heisenberg and E. Heisenberg, *My Dear Li*, 110.

"*resolved to subject*": Heisenberg, *Inner Exile*, 66.

"*if possible, to spare the children*": Richard von Schirach, *Night of the Physicists: Operation Epsilon: Heisenberg, Hahn, Weizsäcker and the German Bomb* (London: Haus, 2015), 7.

"*the soil was rocky*": Heisenberg, *Inner Exile*, 94.

"*This life of an industrial tycoon*": W. Heisenberg and E. Heisenberg, *My Dear Li*, 107.

"*the tendency of uranium*": Bo Lindell, *The Sword of Damocles: The History of Radiation, Radioactivity, and Radiological Protection, Part 2: The 1940s* (Nordic Society for Radiation Protection: 1999), 134.

22. Copenhagen

"*knew that golden bridges*": Heisenberg, *Inner Exile*, 79.

23. 1942

"by no means encouraging": Albert Speer, *Inside the Third Reich: Memoirs* (New York: Macmillan, 1970), 226.

"push fantastic projects": Speer, 226.

"Lenard had instilled the idea": Speer, 228.

"energy-producing uranium motor": Speer, 227.

"he retained control of the ongoing atomic research": W. Heisenberg and E. Heisenberg, *My Dear Li*, xiii.

"still mainly justifying this": Lindell, *Sword of Damocles*, 138.

24. War in the Service of Science

"an excellent experimenter": Goudsmit, *Alsos*, 121.

"now they used the slogan": Goudsmit, 157.

"The work at the institute is also progressing": W. Heisenberg and E. Heisenberg, *My Dear Li*, 230.

"Speleological Research Institute": Jim Baggot, *The First War of Physics: The Secret History of the Atom Bomb 1939–1940* (New York: Pegasus Books, 2010), 271.

"In our nuclear physics group": W. Heisenberg and E. Heisenberg, *My Dear Li*, 239.

25. Building B-VIII

"lovely, large room": Heisenberg, *Inner Exile*, 93.

"he was no longer striving": Heisenberg, 94.

"Although I am getting something to eat": W. Heisenberg and E. Heisenberg, *My Dear Li*, 249.

"It may, perhaps, be crazy": W. Heisenberg and E. Heisenberg, 241.

"The apparatus was still": Werner Heisenberg, *Nuclear Physics* (New York: Philosophical Library, 1953), 170.

"Can you be my witness?": W. Heisenberg and E. Heisenberg, *My Dear Li*, 252.

26. Farm Hall

"Everything was beginning to disintegrate": Heisenberg, *Inner Exile*, 105.

"And finally, suddenly": Heisenberg, 105.

"fetched the last bottle": Heisenberg, 106.

"had perhaps lived too long under the myths": Goudsmit, *Alsos*, 120.

"If American colleagues wish": Goudsmit, 113.

"Microphones installed?": Bernstein, *Hitler's Uranium Club*, 78.

"Detained since more than half a year": Bernstein, foreword.

"with the help of considerable alcoholic": Bernstein, 115.

"All I can suggest": Bernstein, 116.

"If the Americans have a uranium": Bernstein, 116.

"made failure into a virtue": Transcript of interview with Edwin Seaver on Alsos, 1947, Samuel A. Goudsmit papers, Series 01 Biographical Materials box 1, folder 4, Niels Bohr Library and Archives.

"power stations, stations for": Heisenberg, *Nuclear Physics*, 171.

"You all have worked for Nazi Germany": Lise Meitner, "Letter to Otto Hahn [June 27, 1945]," in *Physics and National Socialism*, 333.

"In the dim light": Goudsmit, *Alsos*, 127.

27. The 400

"was obtained by the U.S. Military": Memorandum: French and British Representative, SRD, from US Representative, September 28, 1954, NARA II, RG 466, folder: Disposal of Uranium Cubes 13.

"material cannot contain": To MA Italy, from CSGID Intelligence Division, July 16, 1949, NARA II, RG 59, country files: 35, Germany Federal Republic m. Uranium Cubes.

"Product of German Atomic Pile": To MA Italy, from MILATTCHE Rome, Italy, July 16, 1949, NARA II, RG 59, country files: 35. Germany Federal Republic m. Uranium Cubes.

"At least one block": From J. A. Boatright, Office of the District Attorney, to David Lilienthal, Chairman AEC, June 14, 1949, NARA II, RG 59, country files: 35. Germany Federal Republic m. Uranium Cubes.

"a friend of true democracy": To Harry S. Truman, from "A Friend of True Democracy," April 3, 1951, NARA II, RG 59, country files: 35. Germany Federal Republic m. Uranium Cubes.

"The Commission is aware": From J. K. Gustafson to Mr. R. Gordon Arneson, Department of State, April 26, 1949, NARA II, RG 59, country files: 35. Germany Federal Republic m. Uranium Cubes.

"It appears that the members": From Joseph Case, Office of the Secretary, to Dr. Walter Colby, AEC, March 16, 1951, NARA II, RG 59, country files: 35. Germany Federal Republic m. Uranium Cubes.

"According to the accused": "German Couple Charged with Having Uranium Block," *Keystone Press Agency*, December 12, 1952.

"In the name of Professor Dr. Otto Hahn": From Dr. Forstmann to Operating Facilities Division HICOG, February 20, 1953, NARA II, RG 466, folder: Disposal of Uranium Cubes 13.

cube in question had been "stolen": From Von Laue to Operating Facilities Division HICOG, February 21, 1953, NARA II, RG 466, folder: Disposal of Uranium Cubes 13.

"This matter has been discussed": From Stephen Brown to CA Wendel, April 29, 1953, NARA II, RG 466, folder: Disposal of Uranium Cubes 13.

"We read with interest": From Gordon Arneson to James Penfield, US Embassy London, July 31, 1953, NARA II RG 59, country files: 35. Germany Federal Republic m. Uranium Cubes.

28. Paperweights

"Now let us assume": Goudsmit, *Alsos*, 123.

"But this is uranium!": Goudsmit, 123.

"It was too bad": Goudsmit, 28.

"the type of experiments": Memorandum: Activity of the Material, from S. Goudsmit to Major Furman, May 21, 1945, NARA II, RG 200, box 7, folder: Misc. Intelligence.

"About one and a half tons": Groves, *Now It Can Be Told*, 242.

30. The Race

"there was every reason": Compton, *Atomic Quest*, 7.

"If such a weapon": Compton, 8.

"For us the fight had now begun": Compton, 9.

"Though in one sense competing": Compton, 78.

"Obviously there would be no point": H. D. Smyth, *A General Account of the Development of Methods of Using Atomic Energy for Military Purposes Under the Auspices of the United States Government: 1940–1945* (Washington, DC: US Government Printing Office, 1945), 65.

31. Belgian Uranium

"There is not a grotesque": Arthur Conan Doyle, *The Crime of the Congo* (Sandman Books, 2020), 3.

"The royal palace of Belgium": Mark Twain, *Autobiography of Mark Twain*, vol. 2, eds. Benjamin Griffin and Harriet Elinor Smith (Oakland, CA: University of California Press, 2013), 307.

"I've been waiting for you": Compton, *Atomic Quest*, 96.

32. The Combined Development Trust

"*We wanted to be certain*": Groves, *Now It Can Be Told*, 170.
"*He might as well have tried to hoard*": Rhodes, *Making of the Atomic Bomb*, 500.

33. Murray Hill

"*the story of the supply of uranium*": Compton, *Atomic Quest*, 90.
"*a man who was experienced*": Groves, *Now It Can Be Told*, 181
"*The president has tasked the Corps*": Patricia Guarin-Roper, author interview, October 14, 2020.
"*a remarkably thorough knowledge*": Groves, *Now It Can Be Told*, 182.

34. Making Metal

"*All I need is a lump*": Compton, *Atomic Quest*, 90.
"*to us . . . every week was important*": Compton, 92.
"*code names such as Biscuit*": Larry Williams, "Legacy of the Bomb, St. Louis Nuclear Waste," *Saint Louis Post-Dispatch*, February 12–19, 1989, 4.

35. The Last Stop

"*rigged up a rope*": Patricia Walters, author interview, January 20, 2020.
"*That didn't take long*": Tim Koeth, author interview, March 6, 2022.

37. Science in Context

"*These same mistakes*": Samuel A. Goudsmit, "Nazis' Atomic Secrets. The Chief of a Top-Secret U.S. Wartime Mission Tells How and Why German Science Failed in the International Race to Produce the Bomb [October 20, 1947]," in *Physics and National Socialism*, 390.

INDEX

Page numbers in *italics* refer to images.